Contents

GW00703447

Introduction

These Revision Notes concentrate on topics which are common to all Physics syllabuses (A and AS, linear and modular) and include extra topics which are in most syllabuses.

You are likely to be following a syllabus with one or more optional Special Topics (such as Medical Physics, Particle Physics or Astrophysics). This Guide does *not cover* these options although obviously they are based on the Physics covered here.

It is therefore vital that you have your own copy of the syllabus you are using. An official syllabus may use terms you aren't familiar with – your teacher will help you interpret it. There are a few areas where different textbooks and syllabuses use slightly different conventions to the ones here, often resulting in a difference of sign. They are highlighted in the margin notes, but watch out for them – use this as a work book and write in it: that's why you've been left lots of space!

Two general themes cutting across all numerical work in Physics can cause problems until you've got your mind round them (and penalty marks in an exam if you haven't). These are *SI units* and *significant figures*; this book addresses these thoroughly, in the worked examples in the text and in the answers to questions at the end.

Units are there to help and guide you, not just to please an examiner. Knowing the units of a quantity will remind you about it: if you know density is measured in kg m^{-3} it can *only be* mass divided by volume. When substituting numerical values into a relationship *always* put the SI unit in next to it, not as an afterthought. You will be expected to know the common multiples and sub-multiples: kilo (k), mega (M), giga(G); milli(m), micro(μ), and nano(n). If numerical data is quoted using one of these, the safest way of getting the right answer is to convert the quantity to the base SI unit with the correct power of 10 as early as possible: this is particularly important for areas and volumes when dimensions are given in mm. An algebraic symbol used for a physical quantity includes within it the unit as well as the numerical value. So for example the symbol 'v' in a kinetic energy calculation stands for, say, '10 ms^{-1}' not just 10.

A simple rule for significant figures is to quote a final answer to the same number of significant figures as the original data – if the data has a mixture of significant figures, take the smallest number. Be wary of rounding answers too soon in a calculation: you can end up losing important information!

Finally, you are likely to have the use of a formula sheet in the exam. Make sure you know what is on it, and what isn't. Ideally for every relationship quoted you need to know its status and ask yourself:
- When can I use it? (Always or only in certain situations?)
- Does it summarise an experimental result? (What do I have to know about the experiment?)
- Is it a theoretical result? (Could it be derived? Do I have to do it?)

Apart from this . . . Good Luck.

1 Motion and Forces

Motion in a straight line

Three terms are used to describe motion in addition to the time t as an independent variable:
- distance travelled x
- velocity v
- acceleration a

Velocity is:
- rate of change of distance with time (unit: m s^{-1}) and so is
- gradient of the $x - t$ graph, $\Delta x/\Delta t$

Acceleration is:
- rate of change of velocity with time (unit: m s^{-2}) and so is
- gradient of $v - t$ graph, $\Delta v/\Delta t$

Distance travelled is also:
- area under the $v - t$ graph

Uniform acceleration

Graphs for uniform acceleration

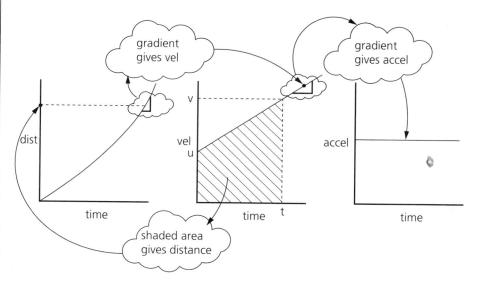

> Caution! only for uniform acceleration.

```
Equations for uniform acceleration
        v = u + at
        x = ut + 1/2 at²
        v² = u² + 2ax
```

$$v = u + at$$
$$x = ut + \tfrac{1}{2} at^2$$
$$v^2 = u^2 + 2ax$$

Motion under gravity (in a vacuum)

- Is described as *free fall*.

Acceleration is:
- constant
- the same for all objects
- denoted by g

Taking height *increasing* (upwards) as positive:

$$g = -9.8 \text{ m s}^{-2}$$
on the Earth's surface
(varying slightly according to location)

Velocity–time graph for an object launched vertically upwards

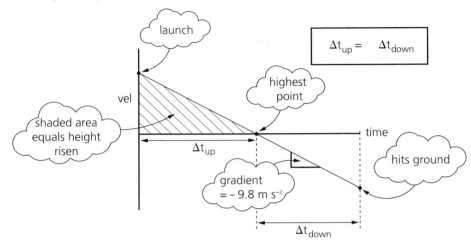

Vectors

- Quantities having magnitude *and* direction.
- If *either* changes the vector changes.

They are normally printed in **bold type** – when handwritten they are usually <u>underlined</u>, e.g. **v** and <u>v</u>; if v is written with neither marking it denotes the magnitude only – a *scalar* quantity.

Vectors for describing motion:
- displacement, **x** – distance from an origin in a *specific direction*
- velocity **v** – the magnitude v is the (scalar) speed
- acceleration **a**.

It is still true that $\mathbf{v} = \Delta\mathbf{x}/\Delta t$ and $\mathbf{a} = \Delta\mathbf{v}/\Delta t$ but in working out the changes $\Delta\mathbf{x}$ and $\Delta\mathbf{v}$ the change in direction needs to be taken into account as well.

Combining vectors

Any vector physical quantity can be represented as:
- a line drawn to some suitable scale
- in the appropriate direction.

Vector addition

To find the effect of many vectors acting together (*vector addition*) the lines representing them are joined head-to-tail

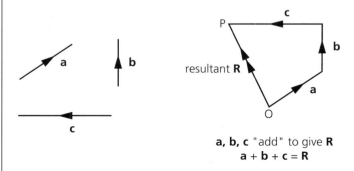

a, b, c "add" to give **R**
a + **b** + **c** = **R**

- It does not matter in which order they are drawn.
- The vector **OP** is called their *resultant* and is the *single* vector which *replaces* the separate ones.
- It can happen that the starting and finishing points coincide so that the resultant is zero.

Special case of 2 vectors at 90°

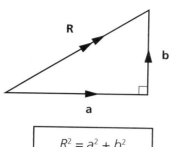

$$R^2 = a^2 + b^2$$
$$R = \sqrt{(a^2 + b^2)}$$

Resolving into components

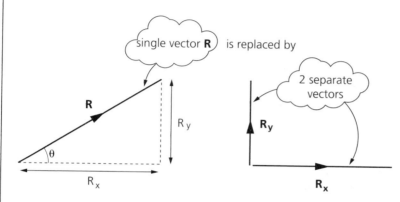

single vector **R** is replaced by

2 separate vectors

- A vector **R** at an angle θ to a given direction can be split up into 2 *component* vectors.
- **R$_x$** along the given direction and **R$_y$** perpendicular to it:

$$R_x = R \cos\theta$$
$$R_y = R \sin\theta$$

- The relationship can work the other way to find θ if the components are known:

$$\tan\theta = R_y/R_x$$
$$\theta = \tan^{-1}(R_y/R_x)$$

Example

The pointer on a computer screen follows *exactly* the motion of the mouse relative to the mat. If the mouse is moved sideways (horizontally on the screen) at a speed of 10 cm s^{-1} and the mat slips at 3 cm s^{-1} perpendicular to this (both velocities relative to the table) how is the velocity vector **v** of the pointer on the screen described?

$$v_x = 3 \text{ cm s}^{-1}$$
$$v_y = 10 \text{ cm s}^{-1}$$
$$v^2 = v_x^2 + v_y^2$$
$$v^2 = 9 \text{ cm}^2 \text{ s}^{-2} + 100 \text{ cm}^2 \text{ s}^{-2}$$
$$v = 10.4 \text{ cm s}^{-1}$$

If θ is the angle of the path to the horizontal $\tan\theta = v_y/v_x = 10/3$ $\theta = 73$ degrees	v is 10.4 cm s^{-1} at 73 degrees to the horizontal

Motion in a curve

If motion is in a *curved* path the direction of **v** is changing and there must be an acceleration (even if the *speed v* is constant).

Projectile under gravity

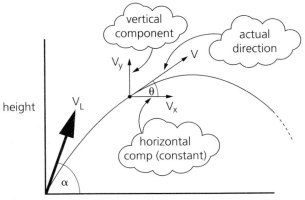

- The shape of the path is a parabola.
- The direction of the velocity **v** at any instant is along the tangent to the curved path at angle θ to the horizontal.

The velocity can be resolved into components:
- horizontal, **v$_x$**, which is constant
- vertical, **v$_y$**, which follows the normal pattern for vertical motion under gravity, acceleration 9.8 m s^{-2} downwards.

If the launch velocity is **V$_L$** at an angle α to the horizontal:

$$v_x = V_L \cos\alpha$$
and is constant

Circular motion at uniform speed

angular velocity ω

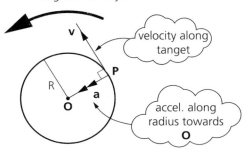

- Direction of the velocity vector is changing.
- So there must be an acceleration.
- Which is directed *towards* the centre (*centripetal*).
- The velocity and acceleration vectors are therefore always at 90° to each other.

It is sometimes convenient (particularly when *time* for a revolution is involved) to work with the *angular velocity* ω (radian per second).

- This is the rate at which the radius vector **OP** rotates; $\omega = v/R$.

> magnitude of
> centripetal acceleration is
> v^2/R or $\omega^2 R$

Newton's Laws of Motion

Second Law

- The effect of a force is to produce a constant acceleration when it acts on a fixed mass.
- The relationship between force **F** (the resultant if more than one), mass m and acceleration **a** is called *Newton's Second Law*:

  F = m**a** (unit: kg m s⁻² or newton, N)

- Since **a** is a vector, so is force.
- The **weight** of an object is the force of gravity on it and must be measured in newtons.
- Since in free fall **a** = **g** an object of mass m must have a weight **W** given by:

  W = m**g**

- **g** is also called the *gravitational field strength* and can be measured in N kg⁻¹ (for more on this see Chap. 6).

First Law

If **F** is zero it follows that **a** must be. This is a version of *Newton's First Law*:

> If there is no resultant force
> the velocity is constant

> The reasoning can be reversed:
> If the acceleration is zero
> the *resultant* force is zero

An example is the motion of a parachutist falling with constant velocity.

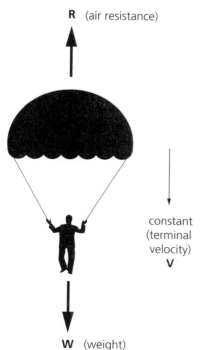

R (air resistance)

constant
(terminal
velocity)
V

W (weight)

The resultant positive (upwards) force **F** is **R** − **W**
Since the acceleration is zero **F** = 0, i.e. **R** = **W**

Third Law

Forces in the universe come in *pairs* of the *same kind*.
Each member of the pair:
- has the same magnitude
- is of the opposite sign to the other
- acts on a different object.

This is a statement of *Newton's Third Law*.
If one member of the pair is known it is not always obvious where its partner is.
- For the parachutist, weight **W** is one of a pair; its partner is the attraction on the Earth of the parachutist, pulling the Earth upwards (effectively acting at the centre of the Earth).
- The Earth is kept in its orbit around the Sun by the gravitational attraction of the Sun on the Earth; the other member of the pair is the force of the Earth on the Sun – both forces are equal and directed *towards* each other.

> Although R and W are equal here they aren't a 'pair' in the 3rd law sense.

Momentum

Change of momentum and impulse

The product *mass × velocity* is called the *momentum*, **p** (a vector):

$$\mathbf{p} = m\,\mathbf{v} \text{ (unit: kg m s}^{-1})$$

The Second Law can be written as:

$$\mathbf{F} = m\,\Delta\mathbf{v}/\Delta t = \Delta(m\,\mathbf{v})/\Delta t$$
(if mass is constant)

or

$$\mathbf{F} = \Delta\mathbf{p}/\Delta t$$
- This gives an alternative wording for Newton's Second Law

> Learn to recognise where this version is more useful – often in continuous flow of mass as in the helicopter example.

Force equals
rate of change of momentum

Example

A helicopter draws air vertically downwards through its blades, the air leaving the blades with a velocity of 30 m s⁻¹. The length of a blade is 3.0 m and the density of air is 1.2 kg m⁻³. What is the upwards lift force on the helicopter?

Stage 1
Calculate the mass of air per second passing through the blades:

> area swept out by blades
> $= \pi\,(3.0 \text{ m})^2 = 28.3 \text{ m}^2$
>
> a cylinder of air 30 m high passes through every second
> so volume per second is
> 28.3 m² × 30 m s⁻¹ = 850 m³ s⁻¹
> mass per second is 850 m³ s⁻¹ × 1.2 kg m⁻³
> = 1020 kg s⁻¹

Stage 2
Calculate momentum per second gained by the air:

> assuming the air well above the blades has no vertical momentum,
> $\Delta\mathbf{p}/\Delta t = 1020 \text{ kg s}^{-1} \times 30 \text{ m s}^{-1}$
> = 3.1 × 10⁴ kg m s⁻² (or N)

Stage 3
Interpret the result:

> The force of 3.1×10^4 N is the
> *downwards* force on the air *from the helicopter*
> By the 3rd law it must also be the
> *upwards* force on the helicopter *from the air*

The momentum version of the Second Law can be re-arranged to:

$$\Delta \mathbf{p} = \mathbf{F} \Delta t$$

The product $\mathbf{F}\Delta t$ (force multiplied by a time interval) is called the *impulse* of the force

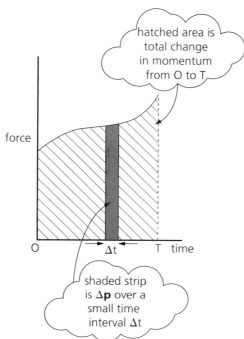

hatched area is total change in momentum from O to T

shaded strip is $\Delta\mathbf{p}$ over a small time interval Δt

> change in momentum equals the area under a force–time graph

Conservation of momentum

- If 2 objects A and B collide, $\Delta\mathbf{p}_A$ for A is $-\Delta\mathbf{p}_B$ for B (since the forces acting are a 3rd Law pair and they act for the same time).
- So their combined momentum stays the same.

This is a simple case of the *Law of Conservation of Momentum* and applies to any number of colliding masses (not just 2).

> Provided no **external** forces are acting
> the total momentum of a system of interacting particles
> remains the same

Example 1
A stationary radium nucleus of mass 3.8×10^{-25} kg emits an α-particle, during a radioactive change, of mass 6.6×10^{-27} kg and velocity 1.2×10^7 m s^{-1}. Explain how the Law of Conservation of Momentum might apply and describe what happens to the resulting radon nucleus.

'Collision' need not mean contact (on an atomic scale what is contact?). A comet passing around the Sun once could be said to be making a collision ('interaction' is better).

Stage 1
Consider whether the law can apply:

> There must be no external force, so for it to be of use the effect of nearby atoms has to be ignored (this may be an invalid assumption)

Stage 2
Apply the law qualitatively:

> The initial momentum before emission is zero, so the final momentum must be zero. So the momentum of the α-part. in one direction is equal in magnitude to the momentum of recoil of the radon nucleus in the other

Stage 3
Write the law as an equation with initial momentum on one side and final on the other. Call the α-particle velocity positive:

> $$0 = \{6.6 \times 10^{-27}\ \text{kg} \times (+1.2 \times 10^7\ \text{m s}^{-1})\} + (\mathbf{v}_{Rn} \times 3.7 \times 10^{-25}\ \text{kg})$$
> $$\mathbf{v}_{Rn} = -2.1 \times 10^5\ \text{m s}^{-1}$$

Stage 4
Interpret the result:

> The radon nucleus has an initial recoil velocity of 2.1×10^5 m s^{-1}

Example 2
A ball of mass 0.3 kg hits the ground with a velocity of 12 m s^{-1} and rebounds with a velocity of 8 m s^{-1}. (a) What is the change in momentum? (b) Why is momentum not conserved?

(a) Decide on a sign for velocity. As with other motion under gravity, take upwards as positive (remember there is no *special* reason for this)

> Final momentum $\mathbf{p}_f = 0.3$ kg $\times$ (+8 m s^{-1})
> $= +2.4$ kg m s^{-1}
> Initial momentum $\mathbf{p}_i = 0.3$ kg $\times$ (−12 m s^{-1})
> $= -3.6$ kg m s^{-1}
> $\Delta\mathbf{p} = \mathbf{p}_f - \mathbf{p}_i$
> $= 2.4$ kg m s^{-1} − (−3.6 kg m s^{-1})
> $= +6.0$ kg m s^{-1}

So there is an *upwards* momentum change of 6.0 kg m s^{-1}; this must be of the right sign since the ground will push the ball upwards.

(b)

> For the ball there *is* an external force – that of the ground – so the law cannot apply here

N.B.
- Even if it bounced back at the *same* speed there would still be a momentum change (of +7.2 kg m s^{-1}).
- For the system ball–Earth the contact force is now *internal* and for the two together momentum *is* conserved.

Equilibrium of forces

Moment of a force

The *moment* is the turning effect of a force about a point

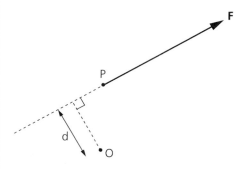

The force **F** acts at P. Its moment about O is:

> **F** *d*
> where *d* is the
> perpendicular distance to the
> line of action of **F** from O

If F acts through O its moment about O is zero.

- Here the moment is clockwise (say positive).
- If O were on the other side of the line of **F** the moment would be negative.

Couple

A pair of forces which are:
- parallel
- equal in magnitude
- acting in opposite directions not through the same point
  is called a *couple*.

A couple is not associated with any particular point.

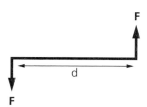

- The combined moment of the 2 forces about *any point* is the same and equals **F**d.
- This is called the moment (or *torque*) of the couple.

Conditions for equilibrium

Any extended object (not just a point mass) acted on by a system of forces in the same plane (*coplanar*) will be in equilibrium if:
- the vector sum of all the forces (the resultant) is zero
- the sum of the moments of all the forces about any point is zero (i.e. clockwise moment equals anticlockwise).

These directions are often horizontal and vertical but they don't have to be – one should be perpendicular to an unknown force.

Procedure for applying conditions

For zero resultant:
- choose 2 perpendicular directions
- resolve all the forces in *each* of these directions
- for each direction, add the components and put equal to zero.

For zero moment:
- choose any convenient point for taking moments about (preferably one through which an unknown force acts)
- calculate the moment of each force about that point (allowing for sign)
- add the moments and put equal to zero.

Example
Diagram (a) shows a simplified version of the forces in the arm when a load is supported in the hand by a horizontal forearm.

Diagram (b) shows just the forces, in an even more stripped down version where the forearm is considered as a rod with a fixed smooth pivot at the elbow E. The forces and distances are shown for a 7 kg load in the hand. F_e is the force acting on the elbow joint from the upper arm and F_m is the force from the biceps muscle. What are the values of F_e and F_m?

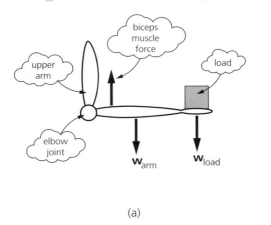

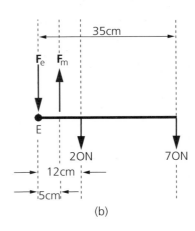

(a)   (b)

Stage 1
Apply condition for zero resultant. Take upwards forces as positive. There are no horizontal forces:

$$F_m + (-F_e) + (-20\,\text{N}) + (-70\,\text{N}) = 0$$
$$F_m - F_e = 90\,\text{N}$$

Stage 2
Apply condition for zero moment. Take moments for each force about the elbow E (since one of the unknown forces acts through E) and call clockwise positive:

clockwise
load: 70 N × 35 cm = +2450 N cm
arm: 20 N × 12 cm = +240 N cm
anticlockwise
muscle: $-(F_m × 5\,\text{cm}) = (-5F_m)$ cm

Stage 3
Total moment is zero:

$$(2450 + 240)\,\text{N cm} + (-5F_m)\,\text{cm} = 0$$
$$5F_m = 2690\,\text{N}$$
$$F_m = 538\,\text{N}$$

Stage 4

Feed this back to the zero resultant condition:

$$538\,\text{N} - \mathbf{F}_e = 90\,\text{N}$$
$$\mathbf{F}_e = 448\,\text{N}$$

Notice that the muscle and elbow joint forces are equivalent to supporting masses of between 45 and 55 kilogrammes!

Work, energy and power

Work done by a force

- Work is done when a force moves its point of application through a distance which is *not perpendicular* to the line of the force.
- The value of the work is equal to:
  force × distance moved *in the line of action of the force*
- If force and displacement are not in the same direction, the force must first be resolved in the direction of the displacement, and then that component used.

$$W = F \times d \text{ (unit: N m or joule, J)}$$
work is a *scalar quantity* so magnitudes only are shown

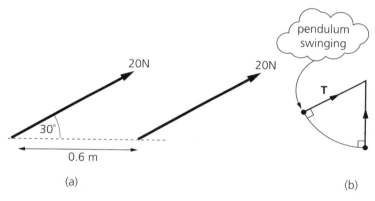

(a)   (b)

- In (a) the work done is:
  $W = (20 \cos 30°)\,\text{N} \times 0.6\,\text{m} = 10.4\,\text{J}$
- In (b) the work done by the tension during the swing is *zero* since it always acts at 90° to the motion.

Energy

Whenever work is done, energy is transferred in one or both of the following ways:
- from one place to another
- from one kind to another.

The value of the energy transfer is *equal to the work done*:
- so energy has to be measured in work units – joule
- energy is a scalar quantity – no direction is involved.

The term *mechanical energy* covers:
- kinetic energy, E_k
- potential energy, E_p.

Kinetic energy (KE)

- Is possessed by any moving object.
- Equals the work done in giving it the velocity (from rest) or in bringing it back to rest.
- Is given by the relation:

$$E_k = 1/2\ mv^2$$
or for a change in velocity
from u to v
$$\Delta E_k = 1/2\ mv^2 - 1/2\ mu^2$$

Example

In a car impact test a vehicle of mass 1500 kg is driven into a barrier at 30 m s^{-1}. It is effectively brought to rest in a distance of 0.80 m. What is the average retarding force F?

Stage 1
Calculate the KE:

$$E_k = 1/2 \times (1500\ \text{kg}) \times (30\ \text{m s}^{-1})^2$$
$$= 6.8 \times 10^5\ \text{J}$$

Stage 2
Equate the loss in KE to the work done:

$$\Delta E_k = 0 - 6.8 \times 10^5\ \text{J}$$
$$F \times 0.8\ \text{m} = -6.8 \times 10^5\ \text{J}$$
$$F = -8.4 \times 10^5\ \text{N}$$
the negative sign correctly
indicates a retarding force

NB
This problem could be tackled by using $v^2 = u^2 + 2ax$ to find a and then using $F = ma$, but the direct route via an energy change is more efficient if acceleration is not needed explicitly.

Potential energy (PE)

- Is always associated with a particular kind of force (gravity, electric, elastic).
- Is associated with the relative position of objects interacting with the force.
- It does *not depend* on how that position has been reached.
- Only *changes* in PE can be calculated when the force does work, unless a *zero level* (any arbitrary convenient state) has been defined first.

Gravitational potential energy is related to a height change.
(A full but cumbersome notation is $E_{p,grav}$ but if it is obvious that the only form of PE is gravitational it will be shortened to E_g).

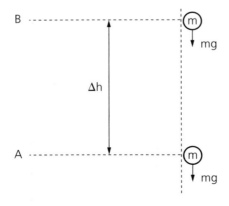

- A mass m is lifted from level A to level B:
  ΔE_g = work done *against* gravity in lifting
  = $(mg)\,\Delta h$

$$\boxed{\Delta E_g = mg\Delta h}$$

- ΔE_g is positive if Δh is positive
  If level A was taken as the zero of gravitational PE *and* heights h were measured from this level *then*:

  at B

$$\boxed{\begin{array}{c} E_g = mgh \\ \text{and the } \Delta \text{ can be dropped} \end{array}}$$

Law of Conservation of Energy

Provided no temperature changes are involved (usually means no friction) the total mechanical energy in a system during any process remains unchanged:

$$\boxed{E_k + E_p = constant}$$

Example

A mass of 2.0 kg is thrown downwards from a height of 5.0 m with a velocity of 4.0 m s^{-1}. What is the total energy during the motion and how fast is it moving when it is 3.0 m above the ground? $g = 9.8$ N kg^{-1} (ignoring its sign here)

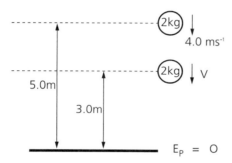

Stage 1
The first part *cannot be answered* until a zero level has been defined – take ground level as $E_p = 0$ (there is nothing inevitable about this choice – it could be *anywhere*).

Stage 2
The total energy is the same everywhere, so take the value at the start:

$$\begin{array}{c} E_{tot} = mgh + 1/2\ mv^2 \\ = (2.0\ \text{kg} \times 9.8\ \text{N kg}^{-1} \times 5.0\ \text{m}) + 1/2 \times 2.0\ \text{kg} \times (4.0\ \text{m s}^{-1})^2 \\ = 114\ \text{J} \end{array}$$

Stage 3
Apply the Law of Conservation of Energy to the second position:

$$\begin{array}{c} (1/2 \times 2.0\ \text{kg} \times v^2) + (2.0\ \text{kg} \times 9.8\ \text{N kg}^{-1} \times 3.0\ \text{m}) = 114\ \text{J} \\ v^2 = (114\ \text{J} - 58.8\ \text{J})/1.0\ \text{kg} \\ = 55.2\ \text{m}^2\ \text{s}^{-2} \\ v = 7.4\ \text{m s}^{-1} \end{array}$$

Elastic and inelastic collisions

- Although during a collision it must be true that the *total energy* remains constant, it is not necessarily true that the total *mechanical* energy remains the same (see *First Law of Thermodynamics* Chap. 2).
- During the actual process of a collision the energy changes may be quite complicated, involving the conversion of at least part of the initial KE to one or more forms of PE (see *elastic potential energy* Chap. 2) and then the recovery of some or all of it back to kinetic towards the end of the impact.
- Nevertheless the overall energy balance can be studied by looking at the total *kinetic* energy before and after the impact, E_i and E_f (no need for a further subscript 'k' since only kinetic is being considered here).

There are 4 possibilities:

- $E_f = E_i$ No loss in overall KE: described as an *elastic collision* (sometimes for emphasis *perfectly* or *totally elastic*)
- $E_f < E_i$ Some loss in overall KE: *inelastic* (or sometimes *partially elastic*)
- $E_f = 0$ All the initial KE is destroyed: described as *totally inelastic*
- $E_f > E_i$ KE is apparently created from nowhere – see the earlier example on the radium nucleus decay: described as *superelastic*.

Caution!
'inelastic'
appears in
both of these.

In the second and third cases you may be asked to discuss possible destinations of the energy *loss* – it ultimately appears as *heat*, often through a variety of intermediate processes.

In the last case you will be looking for a possible *source* of energy – it will ultimately be potential in some form or other (again see, for example, *elastic potential energy* or *electrostatic potential energy* Chap. 6).

However, *whatever category the collision falls into* there is one over-riding principle:

> momentum is *always* conserved
> (with the vital 'no external forces' condition)

Power

This can be thought of as:
either
- the rate of transferring energy

or
- the rate of doing work.

> power = energy transferred/time interval
> $$P = \Delta E/\Delta t$$
> (unit: J s^{-1} or watt, W)

'Rate' means
'per unit time'
(usually per
second)
so divide by
the time
interval.

If a force F moves an object at a *steady speed v* (against some equal resistance force) the power, $P = F v$.

Example
A train delivers a constant power of 2.0 MW when driving at a speed of 60 m s^{-1}. What is the drag force opposing motion?
Call the tractive force of the engine F:

> $$2.0 \times 10^6 \text{ J s}^{-1} = F \times 60 \text{ m s}^{-1}$$
> $$F = 2.0 \times 10^6 \text{ J}/60 \text{ m}$$
> $$F = 33 \text{ kN}$$
> So the drag force is 33 kN in the opposite direction
> since the resultant force on the train must be zero

2   Thermal physics and states of matter

Temperature

The problem is to put a number value to the sensation of *hotness* (which can be very unreliable and subjective).
This is done by:
- deciding what is meant by saying A is hotter than B
  and then
- devising a suitable scale to quantify this difference.

Before getting into the fine points here check with syllabus how much detail is needed.

A scale of temperature

Qualitatively, temperature is that property of an object which determines whether heat flows into it or out of it when placed in contact with another.

- If objects A and B are placed together (into *thermal* contact) and heat flows *from* A *to* B then A is hotter than B: $T_A > T_B$.
- If there is no heat flow between them they are at the same temperature and are in *thermal equilibrium* with each other.

  In order to define the scale (give the temperature a numerical value) the following steps are needed:
- Choose a *thermometric property, X*. This is some property of a material which changes smoothly as its temperature changes (usually, but not necessarily, increasing as temperature increases).
  Examples of X are:
  – the length of a thread of mercury in a glass capillary tube
  – the resistance of a coil of wire
  – the pressure of a constant volume of gas
  – the emf of a thermocouple (a circuit containing 2 junctions of different metals – when they are at different temperatures an emf is produced)
- Choose 2 fixed points as reference points, normally
  – the *ice point* (0 degrees on the X-scale)
  – the *steam point* (100 degrees on the X-scale)
- Measure X at the fixed points, values X_0 and X_{100}

Your textbook will give you precise specification of the fixed points.

- Temperature is assumed to vary linearly with X, so that if X_θ is the value of X at some temperature θ (on the X-scale):

$$\theta/100 = (X_\theta - X_0)/(X_{100} - X_0)$$
$$\text{or}$$
$$\theta = \{(X_\theta - X_0)/(X_{100} - X_0)\} \times 100$$

this is described as the
centigrade temperature on the X-scale

Centigrade just means there are 100 units between the ice and steam points.

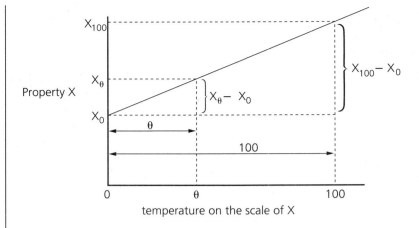

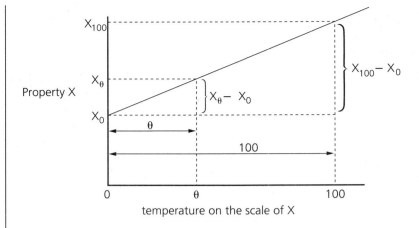

Property X

X_{100}

X_θ

X_0

$X_\theta - X_0$

$X_{100} - X_0$

θ

100

0   θ 100

temperature on the scale of X

An absolute scale

There are two related difficulties with the above process.
- The same temperature measured on two different thermometric scales will not necessarily have the same numerical value (except at the fixed points).
- It is not obvious whether one particular thermometric property X is 'better' than another.

An *absolute scale* overcomes this problem by choosing a thermometric property which is independent of any particular substance.

- The thermometric property is a theoretical one – the *efficiency of an ideal heat engine* (see later).
- So the scale is theoretical (also called a *thermodynamic scale*) and cannot be used practically.

But:

- the temperature scale defined by the *Ideal Gas Equation* (see later) is identical to it, *and in practice*
- the *constant volume gas scale* is very close to it.

The fixed points are:
- absolute zero
- the triple point of water.

The commonest thermodynamic scale is the *Kelvin* scale (SI unit: kelvin, K) where:

> absolute zero is 0 K
> triple point of water is 273.16 K

For everyday purposes the Celsius scale (°C) is used, which is *directly* related to the Kelvin scale:

$T_{celsius} = T_{kelvin} - 273.15$

(The slight difference between 273.16 and 273.15 is to ensure that the ice point is at the familiar 0 °C – and hence that the steam point is at 100 °C: the Celsius scale is a centigrade scale.)

Three points need noting:
- for most purposes, 273 is sufficiently accurate and is commonly used except at the highest precision
- the symbol T (upper case) normally refers to Kelvin temperatures without any further qualification
- even if the Celsius scale is being used, a temperature *interval* $\Delta T_{celsius}$ is given the unit kelvin.

The triple point is the temp. when all 3 phases (solid, liquid, vapour) are in equilibrium with each other.

Note the Celsius scale is defined from the Kelvin.

Heat and energy

Internal energy

Any object above the absolute zero of temperature has an *internal energy U* which is:
- the sum of the random kinetic and potential energies of its individual molecules
- related to the *temperature* and increases as the object becomes hotter
- related to the *state* (solid, liquid or gas) and changes during melting (or freezing) and vaporising (or condensing) even *though the temperature does not change* during these processes
- impossible to measure directly, but changes to it, ΔU, can be measured.

> For a temperature change ΔT of a mass m
> which has a *specific heat capacity c* (unit: $J \, kg^{-1} \, K^{-1}$)
> $$\Delta U = mc\Delta T$$
>
> For a change in state of a mass m
> which has a *specific latent heat L* (unit: $J \, kg^{-1}$)
> $$\Delta U = mL$$

The product mc is called just *heat capacity* (unit: $J \, K^{-1}$)

- For melting ΔU is positive and the specific latent heat of fusion, L_f, is used.
- For vaporisation ΔU is positive and the specific latent heat of vaporisation, L_v, is used.
- For the reverse processes ΔU is negative and of the same magnitude.

First Law of Thermodynamics

This is an extension of the Law of Conservation of Energy to include thermal effects (changes in *temperature* or *state*).

It considers *two* ways in which an object can have its internal energy increased:

- energy Q is transferred *to it* through a temperature difference
- mechanical work W is done *on it* by some external force.

The Law says that the increase in internal energy is the sum of these two quantities.

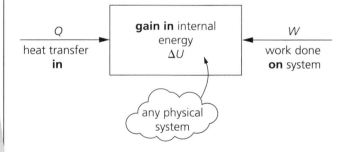

> First Law of Thermodynamics
> $$\Delta U = Q + W$$

- If $Q + W = 0$ ($Q = -W$, heat flow *in* equals work done *by* the system) then $\Delta U = 0$: there is no change in temperature and the change is *isothermal*.
- If $Q = 0$ (complete thermal insulation or not enough time for heat flow) then $\Delta U = W$: an *adiabatic* change.

The term 'heat transfer' Q is reserved for an energy flow driven by a temperature difference.

Care needed here! Some books take W as positive when work is done by the system.

Watch the important little words 'by', 'to', 'on' here – they affect the sign of the terms.

Example

A cylinder fitted with a piston contains 0.1 kg of air and its temperature changes from 30 °C to 430 °C as it expands. The piston exerts a constant force of 10^5 N and pushes a distance of 0.1 m. How much heat energy is transferred into the air? (SHC of air = 1000 J kg^{-1} K^{-1})

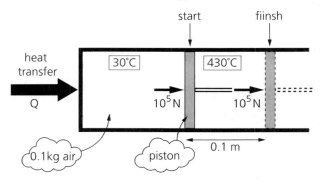

Stage 1

Calculate the external work done:

> The air is expanding so it is doing work *on* its surroundings
> (*W* is *negative*)
> $W = -10^5$ N × 0.1 m
> $= -10$ kJ

Stage 2

Calculate ΔU

> $\Delta U = mc\Delta T$
> $= 0.1$ kg × 1000 J kg^{-1} K^{-1} × (430 − 30) K
> $= +40$ kJ

Stage 3

Apply 1st Law:

> $\Delta U = Q + W$
> 40 kJ $= Q + (-10$ kJ$)$
> $Q = +50$ kJ
> (positive, so heat flow *in*)

Heat engines and efficiency

- A heat engine is any device for converting heat into mechanical energy, i.e. by doing work on its surroundings. It operates in a repeated cycle of processes.
- In its essentials it takes in heat Q_1 from a reservoir at high temperature T_h (*source*) and rejects a smaller amount Q_2 to a reservoir at lower temperature T_c (*sink*), in the process doing work (useful energy output).

The First Law keeps track of the overall energy balance, giving the work done, *W*, as:

$W = Q_1 - Q_2$

But:

- the First Law says nothing about how the input energy Q_1 is distributed between *W* and Q_2

$$Q_1 = Q_2 + W$$

$$\text{Efficiency} = \frac{W}{Q_1} = \frac{Q_1 - Q_2}{Q_1}$$

$$\text{Max. theoretical efficiency} = \frac{T_h - T_c}{T_h} = 1 - \frac{T_c}{T_h}$$

- Ideally Q_2 would be zero so that all the input heat Q_1 does external work (100% efficiency).
- Unfortunately this is not possible and is in conflict with the Second Law of Thermodynamics, one version of which states

> The complete conversion
> of heat energy from a hot object into work
> is not possible

The *efficiency* of any machine or engine is the proportion of energy input which is actually delivered as useful energy output:

> efficiency = useful energy output(work done)/energy input

For a heat engine:

> efficiency = $W/Q_1 = (Q_1 - Q_2)/Q_1$

A consequence of the Second Law is that under certain ideal conditions (never quite realised in practice) the efficiency can be written as:

> efficiency = $(T_h - T_c)/T_h$

- This expression is called the *thermodynamic efficiency* and represents a theoretical upper limit – no amount of clever engineering can exceed this value.
- Real heat engines (power stations, internal combustion engines) will have efficiencies appreciably less than this.

- The efficiency can be increased by raising T_h and/or lowering T_c
- The efficiency can only be 1 (100%) when $T_c = 0$, the absolute zero (which is never quite attainable).

This is the definition of the absolute zero.

Heat transfer

Three mechanisms exist.

Conduction

- The transfer of energy through *any* medium (other than a vacuum) without any mass movement.
- Maintained by a temperature difference.
- On a microscopic scale the transfer is by atomic vibration or (in a metal) by free electron energy exchange.

Temperature gradient is negative in the direction of heat flow (hot→cold).

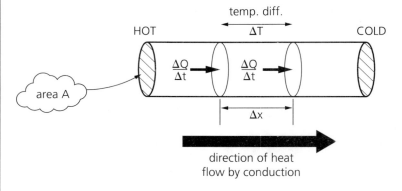

direction of heat
flow by conduction

Provided the surfaces are insulated the rate of heat conduction is the same through any cross-section.

> temperature gradient $\Delta T/\Delta x$ (K m^{-1})
> rate of conduction heat transfer $\Delta Q/\Delta t$ (W)
>
> $\Delta Q/\Delta t = -kA\ (\Delta T/\Delta x)$
> where k is the
> *thermal conductivity* (unit: W m^{-1} K^{-1})

Convection

- Only occurs in fluids.
- Temperature differences cause density differences which produce mass flow.
- Heat transfer is due to the flow of the cold and hot parts and then mixing.

Radiation

- Transfer of energy by electromagnetic waves emitted by hot objects.
- Can therefore be transferred through a vacuum.
- The full range of the spectrum contributes.

But:

Your syllabus may require knowledge of Stefan's and Wien's Laws.

- the contribution from each part of the spectrum (i.e. narrow band of wavelengths) depends upon the temperature of the radiating surface.

Heat loss through building materials

When considering insulation and heat loss in buildings, U-values are often used (unit: W m^{-2} K^{-1})

Not to be confused with U for internal energy!

> $\Delta Q/\Delta t = U\text{-value} \times A \times \Delta t$

- The U-value can be interpreted as the overall heat transfer rate through a material per unit area for each 1 K temperature difference across the material.
- Its value takes in the effect of *all* heat transfer processes.
- Where conduction is the only process, U-value = $k/\Delta x$ and so (unlike k which is a material property) it depends on thickness.

Gases

The Mole

In comparing one gas with another it is often convenient to work in *molar quantities*.
- The *mole* (mol) is the SI unit of amount of substance:

> It is the amount of substance that contains
> as many particles as there are
> carbon atoms in exactly 12 grammes
> of the ^{12}C isotope

- The number of particles in 1 mole is called the *Avogadro constant*, N_A:
$N_A = 6.02 \times 10^{23}$ mol^{-1}
In 12 grammes of ^{12}C there are 6.02×10^{23} atoms of the element.

Molar mass and relative mass

- The *molar mass* of a substance is the mass of one mole of it.
- The *relative mass* of a substance is the mass of a molecule of it compared with one-twelfth of the mass of an atom of the ^{12}C isotope.

The molar mass M in *grammes* is numerically equal to the relative mass:

> In a substance of molar mass M and mass m
> there is an amount n moles where
> $$n = m/M$$

Two cautionary points!
Examination data may be given to you in one of two ways:
- molar mass M in proper SI units of kilogramme per mole (kg mol^{-1})
- relative mass, which is a pure number without any units.

Note that the relative mass gives the molar mass in grammes! much care needed here.

Gas laws

Observable (*macroscopic*) properties of a gas are described in terms of:
- pressure, p, the force exerted per unit area (unit: N m^{-2} or pascal, Pa)
- volume, V (unit: m^3)
- kelvin (or absolute) temperature T (unit: kelvin, K).

There are 2 experimental relationships (which most gases follow fairly well) which for a *constant mass* are:
- Boyle's Law $p \propto (1/V)$ T constant
- Pressure Law $p \propto T$ V constant

You need to be familiar with the graphs of p against V and p against T.

If the mass is allowed to vary as well, it is found that all gases obey the *same* equation provided *number of moles* are used.

The Ideal Gas Equation
$$pV = nRT$$
where n is the number of moles
R is called the Ideal Gas Constant
$R = 8.31$ J mol^{-1} kg^{-1}

- An *ideal gas* is a theoretical model which follows this equation exactly.
- Real gases follow it approximately, particularly at low density and far away from condensing to a liquid.

Kinetic theory model of a gas

Basic assumptions

- A gas consists of a very large number of molecules.
- Molecules are in continuous random motion.
- Molecular volume is negligible in comparison with the volume they occupy.
- No attractive forces between the molecules.
- All molecular collisions are *elastic*.
- The time spent on a collision is very short compared with the time between collisions.

The model:
- explains gas *pressure* by considering collisions with the container walls
- *calculates* the pressure by considering *momentum changes* at a collision
- enables molecular *speeds* to be estimated.

The mean (average) *velocity* of all the molecules in a stationary gas must be zero – however this will not be true for the (scalar) *speed*.
- The *mean square speed* is the average of the squares of individual molecular speeds.
- The *root mean square* (RMS) speed is the square root of this value.
- The kinetic theory deals with the RMS value.

If c is the *root mean square* (RMS) velocity
of N molecules each of mass m
in a volume V
$$pV = \tfrac{1}{3}(Nmc^2)$$

which leads to 2 further
fundamental results:

$$p = \tfrac{1}{3}\rho c^2$$
where ρ is the density

and

$$E_k = \tfrac{3}{2}RT$$
where E_k is the
total KE of random motion
of 1 mole
or *for one molecule*
$$= \tfrac{3}{2}(R/N_A)T$$
$$= \tfrac{3}{2}(kT)$$
where $k\ (=R/N_A)$ is
Boltzmann's constant (1.38×10^{-23} J K^{-1})

Example 1

What is the RMS speed of air molecules at room temperature? Air has a density of 1.2 kg m^{-3} and a pressure of 1.0×10^5 Pa:

$$p = 1/3 \; \rho c^2$$
$$c = \sqrt{(3p/\rho)}$$
$$= \sqrt{(3 \times 1.0 \times 10^5 \text{ N m}^{-2}/1.2 \text{ kg m}^{-3})}$$
$$= 500 \text{ m s}^{-1}$$

Example 2

0.30 kg of argon is at a pressure of 1.0×10^6 Pa in a cylinder of volume 0.02 m^3. The molar mass of argon is 0.04 kg mol^{-1}. (a) How many moles of argon are there? (b) What is the temperature? (c) What is the mean KE of one molecule?
Boltzmann's constant, $k = 1.4 \times 10^{-23}$ J K^{-1}

Stage 1

$$n = m/M = 0.3 \text{ kg}/0.04 \text{ kg mol}^{-1} = 7.5 \text{ mol}$$

Stage 2

Re-arrange the Ideal Gas Equation to give
$$T = pV/nR$$
$$= 1.0 \times 10^6 \text{ Pa} \times 0.02 \text{ m}^3/(7.5 \text{ mol} \times 8.3 \text{ J mol}^{-1} \text{ K}^{-1})$$
$$= 320 \text{ K}$$

Stage 3

mean KE of 1 molecule $= 3/2(kT)$
$$= 1.5 \times 1.4 \times 10^{-23} \text{ J K}^{-1} \times 320 \text{ K}$$
$$= 6.7 \times 10^{-21} \text{ J}$$

A check could be carried out to see how this all ties together by finding the RMS speed in 2 different ways:

A. Find the mass of 1 molecule:

$$m = 0.04 \text{ kg mol}^{-1}/6.0 \times 10^{23} \text{ mol}^{-1}$$
$$= 6.7 \times 10^{-26} \text{ kg}$$

and using the KE formula:

$$(1/2) \times 6.7 \times 10^{-26} \text{ kg} \times c^2$$
$$= 6.7 \times 10^{-21} \text{ J}$$
$$c^2 = 2.0 \times 10^5 \text{ (m s}^{-1})^2$$
$$c = 450 \text{ m s}^{-1}$$

B. Find the density and use the method of example 1:

$$\rho = 0.30 \text{ kg}/0.02 \text{ m}^3 = 15 \text{ kg m}^{-3}$$
$$c = \sqrt{(3p/\rho)} = \sqrt{(3 \times 1.0 \times 10^6 \text{ Pa}/15 \text{ kg m}^{-3})}$$
$$= 450 \text{ m s}^{-1}$$

Properties of solids

Behaviour under load

When a force is applied along the length of a solid of uniform cross-section:
- there is always some deformation
- this may be an increase in length under tension or a decrease under compression.

When the force is removed, the deformation may:
- return to zero – elastic behaviour
- be permanent – inelastic or plastic behaviour.

To compare the behaviour of different materials independently of their size:
- *Stress* (σ) is used rather than force
- *Strain* (ε) is used rather than actual deformation.

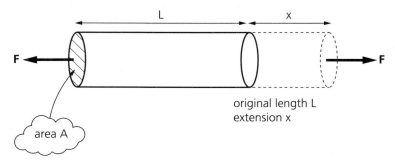

original length L
extension x

area A

$$\sigma = F/A \text{ (unit: N m}^{-2} \text{ or Pa)}$$
$$\varepsilon = x/L \text{ (a ratio – no units)}$$

The behaviour of a particular material is shown by a stress–strain graph.

A typical σ–ε graph for a metal (e.g. copper)

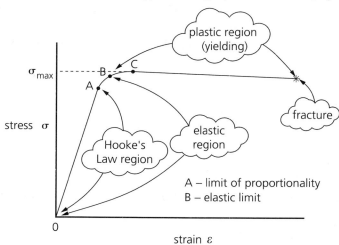

plastic region (yielding)

σ_{max}

stress σ

Hooke's Law region

elastic region

fracture

A – limit of proportionality
B – elastic limit

0

strain ε

- From 0 to A Hooke's Law is obeyed and the graph is linear.
- In this region *stress α strain* or stress/strain is a *constant* – a property of the material called *Young's modulus E* (unit: as for stress, Pa):

$\sigma/\varepsilon = E$

- From A to B the deformation is still elastic but does not obey Hooke's Law.
- From B to fracture the deformation is plastic.
- At C the maximum stress is reached (the ultimate tensile stress).

Relation to microstructure

Not all materials show all the features of the graph shown and some may show more.

How a given material behaves depends on the arrangement and properties of its molecules. There are 2 broad categories of solid structure:
- crystalline
- amorphous.

Crystalline

- All metals are crystalline.
- The atoms are arranged regularly in a *lattice*.
- A metal is often *polycrystalline* – consists of a large number of very small crystals (grains) arranged randomly.
- There are often faults in the crystal lattice called *dislocations* which can move around and can dramatically weaken the metal.

Amorphous

- No regular atomic arrangement over long distances (on an atomic scale).
- Examples are glass and unstretched rubber (and liquids).

> The class of material determines its elastic and plastic behaviour

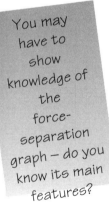

You may have to show knowledge of the force-separation graph – do you know its main features?

Elastic behaviour

- The interatomic bonds are themselves stretched.
- For small deformations the force-separation graph for a pair of atoms or molecules is linear, leading to Hooke's Law.
- All materials show *some* elastic behaviour and have a linear region of the stress–strain graph although it may be very small.

Plastic behaviour

- Caused by layers of atoms sliding over each other
  and
- by movement of dislocations
  but
- not shown appreciably by those amorphous substances which are *brittle*.

Polymers

A class of material which may be crystalline or amorphous.
- Have long chain molecules often coiled up and tangled together.
- May change from amorphous to crystalline depending on the stress.
- If amorphous, are likely to be brittle.
- Can show very large plastic strains, a very small elastic region and virtually no Hooke's Law behaviour.

Here are two further stress–strain graphs which you should be familiar with. They are both for amorphous materials but are different from each other.

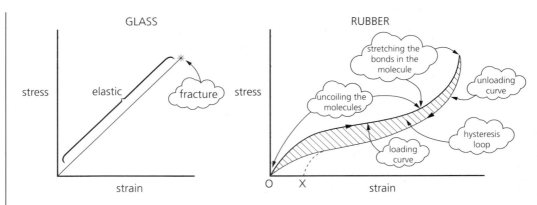

Glass

- Typical brittle behaviour.
- No plastic region.
- Is elastic and obeys Hooke's Law *up to fracture*.
- In principle is very strong but in practice will break by *crack propagation* before its ultimate tensile stress is reached.

Rubber

- Two distinct regions:
  - one of low stiffness (but not constant – it doesn't obey Hooke's Law) where the molecules are uncoiling
  - followed by a much stiffer region where the bonds in the molecule itself are being stretched.
- Capable of much larger strains than most other substances – of the order of 6 compared with copper of around 0.1 at fracture.
- The unloading graph does not retrace the loading graph, showing *hysteresis*.
- The area of the hysteresis loop represents the energy lost during one load–unload cycle (see *elastic potential energy*).
- The unloading process can result in a strain OX at zero stress but this is not a proper plastic strain: it is time-dependent and gradually reduces to zero under no load.

Elastic potential energy

- If a *particular sample* of a material which obeys Hooke's Law is being studied it is convenient to simplify the law to:

> $F = kx$ where k is a constant
> called the *stiffness* (unit: N m^{-1})

- When the sample is stretched the work done on it (converted to *elastic potential* energy E_{el}) is represented as the area under the $F – X$ graph:

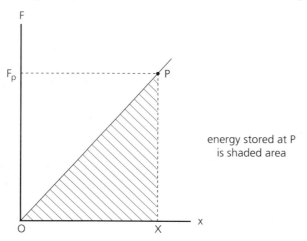

energy stored at P is shaded area

Spring behaviour can usefully be written in this way.

k can also be called the force or spring constant.

The energy stored is always the area under the force-extension graph whatever its shape.

At P the extension is X and the force is F_p:

> The energy at P is the shaded triangle
> $E_{el} = 1/2 \, F_p X$
> or since $F = kx$
> $E_{el} = 1/2 \, kX^2$

Example

A steel rod of length 1.5 m and diameter 5.0 mm is put under a tension of 1.0×10^4 N. Assuming this is still in the Hooke's Law region and Young's modulus is 2.0×10^{11} Pa, how much does it stretch by and how much energy is stored in it?

Stage 1
Calculate stress:

> $$\sigma = F/A$$
> $$= 1.0 \times 10^4 \, N/\{\pi \times (2.5 \times 10^{-3} \, m)^2\}$$
> $$= 5.1 \times 10^8 \, Pa$$

Young's modulus (Y.M.) could also be given as 200 GPa – check you know the prefix G (and M!).

Stage 2
Calculate the strain using definition of Young's modulus:

> $$\sigma/\varepsilon = E$$
> $$\varepsilon = \sigma/E$$
> $$= 5.1 \times 10^8 \, Pa/2.0 \times 10^{11} \, Pa$$
> $$= 2.6 \times 10^{-3} \, (0.26\%)$$

Stage 3
Find extension from strain:

> $$\varepsilon = x/L$$
> $$x = \varepsilon L$$
> $$= 2.6 \times 10^{-3} \times 1.5 \, m$$
> $$= 3.8 \, mm$$

3.8 comes from the 'unrounded' answer of Stage 2.

Stage 4
Use final values of tension and extension to find the energy:

> $$E_{el} = 1/2 \, (1.0 \times 10^4 \, N \times 3.8 \times 10^{-3} \, m)$$
> $$= 19 \, J$$

Fluids

The term applies to matter which can flow easily – gases and liquids.

On a *microscopic* (molecular) level, *liquid structure* shows some of the properties both of gases:
- no regular arrangement (there may be some *short-range* order)
- molecules move randomly,

and solids:
- molecules much closer than in a gas (liquid and solid densities not very different)
- inter-molecular forces obviously significant in giving a liquid a well-defined volume and surface
- energy needed to break the molecular bonds (during vaporisation – latent heat).

Brownian Motion in liquids.

The term 'liquid crystal' reinforces the link with solids.

Pressure at a depth

Both liquids and gases exert a pressure which acts at right angles to any flat surface in the fluid, whatever the direction of the surface.

- The pressure depends on the depth and is the same at all points at the same depth.
- The pressure is due to the weight of fluid above the particular level.
- The pressure therefore increases with depth.

(Although the pressure of the atmosphere at ground level is ultimately due to molecular bombardment, following the kinetic theory model, it can also be thought of as being due to the weight of a vertical column of air above the Earth – see example 1).

> At a depth h in a fluid of density ρ the pressure is given by $p = \rho g h$

You could make a more realistic estimate of the air's mean density and re-calculate h.

Example 1
Assuming the density of air at different heights remains the same as the ground level value of 1.2 kg m^{-3} (which it obviously doesn't) and that the ground level pressure is 1.0×10^5 Pa, what is the thickness of the atmosphere?

$$h = p/\rho g$$
$$= 1.0 \times 10^5 \text{ N m}^{-2}/(1.2 \text{ kg m}^{-3} \times 9.8 \text{ N kg}^{-1})$$
$$= 8.5 \times 10^3 \text{ m (8.5 km)}$$

There is not of course a 'proper top' to the atmosphere.

It's useful to have a rough idea of density values – that for water is a good starting point (and for air it's about 1000 times smaller).

Example 2
Estimate the increase in pressure in descending to a depth of 5 km in the sea.
Since this is an estimate, calculations need only be to 1 sig. fig. (as in 5 km) and the density taken as 1000 kg m^{-3} (thus ignoring the salt content and slight compressibility of water).

$$\Delta p = \rho g \Delta h$$
$$= 1000 \text{ kg m}^{-3} \times 10 \text{ N kg}^{-1} \times 5000 \text{ m}$$
$$= 5 \times 10^7 \text{ N m}^{-2}$$

Note
- This is about 500 times atmospheric pressure.
- The actual total pressure at this depth will be the value calculated *plus* atmospheric pressure acting on the sea's surface – this is obviously negligible here (501 rounds to 500 to 1 sig. fig.) but should not in general be overlooked.

Upthrust and buoyancy

- If an object is immersed totally in a fluid the pressure at its lower surface is greater than the pressure on its upper.
- So the fluid exerts a net upwards force (the *upthrust*) and the object will apparently weigh less than in a vacuum.
- The upthrust is equal to the weight of the fluid which the object has displaced (true even if it is *not all* immersed).
- If the upthrust is greater than the weight of the object then it will float upwards until sufficient of it is above the surface (to reduce the upthrust) to restore vertical equilibrium.
- An object will float in a fluid if its density is less than or equal to that of the fluid.

Known as Archimedes Principle.

So steel floats on mercury.

3 Current electricity

Current

Current is the movement of electrically charged particles.
These may be:
- electrons – the fundamental particle of *negative* charge
- ions – atoms which have gained or lost one or more electrons and may be of either sign.

Metallic conduction

The structure of a metal on an atomic scale is a crystal lattice of *positive ions* in which free electrons (the *conduction electrons*) move around randomly in the spaces.

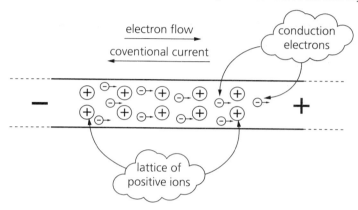

When a battery is connected across the ends of a metal wire:
- an electric field is created in the wire exerting a force on the free electrons
- they drift very slowly towards the positive terminal
- this movement *is* the current
- the *conventional current* (for historical reasons) is in the *opposite direction to electron flow*.

Fluid conduction

This covers current in liquids and gases.
- Only *ionised* (the atomic particles are ions) fluids conduct.
- Some types of liquid are always ionised (*electrolytes* – solutions of salts).
- Gases are not normally ionised at room temperature.
- Gases need a high temperature or high voltage to ionise them.
- Both signs of charge are present, moving in opposite directions.
- The positive ions move in the direction of conventional current.

Unit of current

- Current is one of the 6 base SI quantities and all other electrical quantities derive from it (just as all mechanical quantities are based on mass, length and time).
- The SI unit is the *ampere* (A).
- The definition of the unit is based on the magnetic force between parallel conductors carrying current (see p. 82).

Charge

The amount of charge passing a given place in a circuit is related to:

- the strength of the current, I
- the time for which it flows, Δt.

If ΔQ is the charge:

> $\Delta Q = I\,\Delta t$ (unit: A s or coulomb, C)
> $I = \Delta Q/\Delta t$
> current = rate of flow of charge
> 1 ampere = 1 coulomb per second (1 C s^{-1})

Charge transfers in multiples of the charge on the electron (negative) or proton (positive) which is usually written as e.

> $e = 1.60 \times 10^{-19}$ C

'e' has actually been measured to 8 sig. figs.

Example
A current of 4.0 μA flows for 15 ms. How many electrons pass any particular place?

> Charge passed
> $\Delta Q = 4.0 \times 10^{-6}$ A $\times$ 15 $\times$ 10^{-3} s
> $= 6.0 \times 10^{-8}$ C (60 nC)
> If n electrons pass then
> $ne = \Delta Q$
> $n = 6.0 \times 10^{-8}$ C/1.6×10^{-19} C
> $= 3.8 \times 10^{11}$
> (rounding to 2 sig. figs. as in the data)

You need to know how 'n' and 'v' vary between metals and semiconductors.

Speed of charge carriers
For a conductor of cross-sectional area A containing n charge carriers (electrons or ions) per unit volume each of charge q and moving with mean velocity v, the current flowing is given by:

> $I = nAqv$

Potential difference (p.d.)

- Gives information about electrical energy conversion ΔE between 2 points.
- The conversion is *from* electrical *to* some other form.
- The final form of the energy is heat, but it might be *via* kinetic (motor, accelerating a charge in a vacuum).

Pot. difference always has 2 points associated with it.

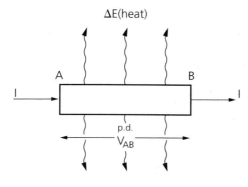

If the charge transported between the 2 points is ΔQ and the p.d. is V:

$$V = \Delta E / \Delta Q$$
(unit: joule/coulomb, J C^{-1} or volt, V)

Example

An electron is accelerated from rest through a p.d. of 15 kV. How fast is it moving? The mass of the electron, $m_e = 9.1 \times 10^{-31}$ kg.

Principle: find the energy converted from the power supply and equate to gain in KE.

Energy converted
$E = V \Delta Q$ and here ΔQ is e
$E = 15 \times 10^3$ V $\times 1.6 \times 10^{-19}$ C
$= 2.4 \times 10^{-15}$ J

$1/2\ mv^2 = E$
$(9.1 \times 10^{-31}$ kg$/2)v^2 = 2.4 \times 10^{-15}$ J
$v = \sqrt{(5.27 \times 10^{15}\ \text{m}^2\ \text{s}^{-2})}$
$= 7.3 \times 10^7$ m s^{-1}

Zero of potential

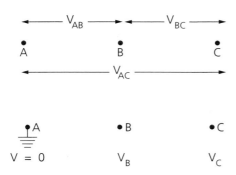

$$V_{BC} = V_{AC} - V_{AB}$$

But, any point can be arbitrarily labelled *zero potential* (just like gravitational PE).
- If A is called zero potential it is written $V_A = 0$ (sometimes called 'earth' or 'ground').
- It is then correct to talk about *the potential* at say B, V_B.
- Can then say $V_{BC} = V_C - V_B$, giving potential difference its proper meaning.

Resistance

Defined by the relation
$R = V/I$
(unit: volt/ampere, V A^{-1} or ohm, Ω)

For any conductor, resistance depends on dimensions.
- Is proportional to the length, ℓ.
- Is proportional to $(1/A)$ where A is the cross-sectional area.

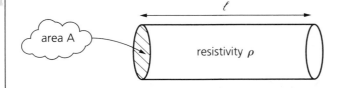

$R \propto \ell/A$
$R = \rho \ell/A$
ρ is the *resistivity* – a property of the
material
(unit: ohm metre, Ω m)

Resistivity varies with temperature: the way it varies relates to the terms n and v in the expression $I = nAqv$.

- For a metal it increases with temperature –
  v is reduced as a result of more electron scattering by atomic vibrations.
- For a *thermistor* (with a negative temperature coefficient, NTC) and other semi-conductors it decreases with increasing temperature: n is increased (by a bigger factor than v is reduced) as more charge carriers are liberated from the atoms.

Ohm's Law

For many conducting materials (mainly metals)
the resistivity is constant at constant temperature

You need to know the I–V graphs for a diode and a lamp filament.

This means that for a *given conductor* made out of an *ohmic* material a graph of current against applied p.d. will be linear through the origin. Whatever the shape of the graph, it is always the case that resistance under particular conditions is the value of V/I at the point concerned.

Note the convention for independent variable on the x-axis.

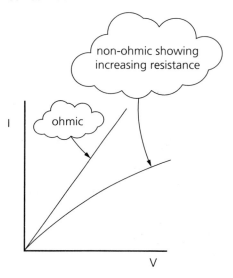

non-ohmic showing increasing resistance

ohmic

Power

- From the defining relation for p.d. an energy transformation ΔE can be written:
  $\Delta E = V\Delta Q$
- If both sides are divided by the time interval Δt it becomes:
  $\Delta E/\Delta t = V\Delta Q/\Delta t$
- The left-hand side is the *power P* (rate of transforming energy) and on the right-hand side the term multiplying V is the *current I*.

$P = VI$
power = p.d. × current

- If this relation is combined with the definition of resistance ($R = V/I$), two other very useful results follow:

$$P = I^2R$$
$$P = V^2/R$$

Some care and judgement is needed in deciding which of the last two to use: when in doubt, go back to $P = VI$ which can *always* be used.

Example 1
A piece of domestic electrical equipment designed to run off the mains has its resistance halved. What happens to the power dissipated?

> Since V is fixed and R varies
> it is sensible to use
> $$P = V^2/R$$
>
> Here $P \propto 1/R$
> so if R halves P will double

Example 2
The mains voltage drops from 240 V to 210 V. How much longer will it take a kettle to boil?

Stage 1
Resistance is fixed (but not known, so call it R) and V is known, so use $P = V^2/R$ again and call the powers P_1 and P_2.

Stage 2

> $P_1 = (240\ V)^2/R$
> $P_2 = (210\ V)^2/R$

Stage 3
Find the ratio of the powers:

> $P_2/P_1 = 210^2/240^2$
> $= (210/240)^2$
> $= 0.77$

Stage 4

> Since the new power is 0.77 (77%) of the old
> the time taken to heat
> (deliver a fixed energy)
> will be *longer* by a factor
> 1/0.77, i.e. 1.31

Stage 5
Final statement:

> Time has increased by 31%

In practical applications the joule is an inconveniently small unit of energy. An alternative in common use is the kilowatt-hour (kW h). This is the energy converted during 1 hour when working at a rate of 1 kW.

1 kW h = 1000 J s^{-1} × 3600 s
= 3.6 MJ

Which to use depends on which of I or V you know.

'How much longer?' expects the answer 'so many times' not an actual time interval.

In ratio problems it is often best to leave the actual evaluation to the end.

Since $P \propto V^2$ you could have written this directly.

Remember energy = power × time.

DC circuits

Power source

Any device for converting *from* another form of energy *to* electrical energy – chemical cell, solar cell, photo-voltaic cell, dynamo.

It is characterised by 2 properties.

- Electromotive force (emf) $\mathscr{E}$ measured in volts.
- Internal resistance, r.

emf = power converted in source/current flowing

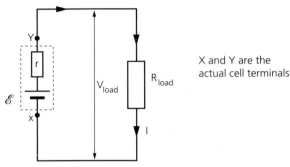

X and Y are the actual cell terminals

- The emf is a property of the source, constant under fixed conditions.
- The internal resistance is the resistance between the terminals *inside* the source.

Complete circuit relation

$$V_{load} = \mathscr{E} - Ir$$
$$\text{where}$$
$$I = \mathscr{E} / (R_{load} + r)$$

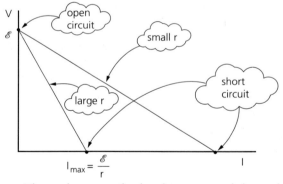

- The p.d. across the load must equal the p.d. across the terminals.
- The graph of load p.d. against current is linear with gradient $-r$ and intercept on the *V*-axis of $\mathscr{E}$.
- The condition of zero current (cutting *V*-axis) is called *open circuit* ($R_{load} = \infty$).
- The condition of maximum current (cutting *I*-axis) is called *short circuit*.

The surprising feature, at first sight, of short circuit is that the terminal p.d. is *zero* (until it is realised that for this to happen R_{load} must also be zero).

Example 1

A car battery has emf 12 V and internal resistance 0.02 Ω. It delivers 200 A to the starter motor. What is the p.d. across the motor and its resistance?

Stage 1
Identify the variables, their numerical values and units:
$\mathscr{E} = 12$ V $I = 200$ A $r = 0.02$ Ω

Stage 2

$$V_{load} = 12\ V - (200\ A \times 0.02\ \Omega)$$
$$= 8\ V$$

Stage 3

$$R_{mot} = 8\ V/200\ A$$
$$= 0.04\ \Omega$$

Example 2

A laboratory high voltage supply has emf of 5.0 kV and must not be capable of delivering more than 3.0 mA for safety reasons. What internal resistance must it have?

Stage 1

Relate the question data to basic properties of a source. 3.0 mA is the maximum current so it must refer to short circuit, i.e. $V_{load} = 0$. Note data is to 2 sig. figs.

Stage 2

Substitute the values into the complete circuit equation:

$$0 = 5000\ V - (3 \times 10^{-3}\ A) \times r$$
$$r = 5000\ V/(3 \times 10^{-3}\ A)$$
$$= 1.7 \times 10^{6}\ \Omega\ (1.7\ M\Omega)$$

Result quoted to same no. of sig. figs. as data.

Maximum power

- A power supply delivers maximum power to a load when:

  load resistance = internal resistance
  the load is said to be *matched* to the source

Under these conditions:
- the load p.d. equals half the emf
- half of the converted power is dissipated as heat in the supply itself
- the other half is the useful power in the load
- so the *overall efficiency* is 50%.

Series and parallel circuits

Series

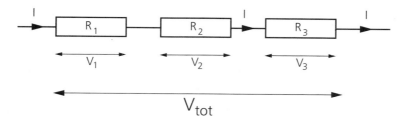

- The current is the same everywhere.
- Potential differences add up: $V_{tot} = V_1 + V_2 + V_3$
- Resistances add up: $R_{tot} = R_1 + R_2 + R_3$

  R_{tot} is the *single equivalent resistance*
  It can replace the separate ones

Parallel

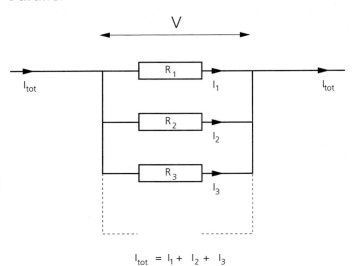

$$I_{tot} = I_1 + I_2 + I_3$$

- The p.d. across each is the *same*.
- The currents in each branch add up to the total flowing in.
- The reciprocals (inverses) of each resistance should be added up:

$$1/R_{tot} = 1/R_1 + 1/R_2 + 1/R_3$$

The combined resistance of a parallel combination is always *smaller than the smallest*.

Example

Find the single equivalent resistance to the parallel combination of 3.7, 4.2 and 5.8 ohm resistances.
The values of R and $1/R$ are set out in the table:

R_1	R_2	R_3	$1/R_1$	$1/R_2$	$1/R_3$	$1/R_{tot}$
Ω	Ω	Ω	Ω^{-1}	Ω^{-1}	Ω^{-1}	Ω^{-1}
3.7	4.2	5.8	0.27	0.24	0.17	0.68

The final stage, which it is easy to forget, is to invert the final value to obtain R_{tot}:

$$R_{tot} = 1/0.68 \ \Omega^{-1} = 1.5 \ \Omega$$

Check: The value is less than 3.7 so is at least sensible!

Potential divider

A circuit consisting of 2 series resistors designed to deliver a proportion of the total p.d. across them to another circuit.

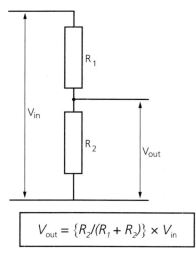

$$V_{out} = \{R_2/(R_1 + R_2)\} \times V_{in}$$

Two useful special cases:
- as $R_2 \to 0$, $V_{out} \to 0$
- when $R_2 \gg R_1$, $V_{out} \to V_{in}$

Example

An LDR is connected into the circuit below. In the dark its resistance is 5000 Ω and in the light 200 Ω. What are the dark and light values of the output voltage?

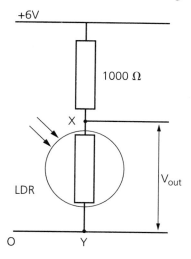

In the dark
$$V_{out} = \{5000/(5000 + 1000)\} \times 6 \text{ V}$$
$$= 5 \text{ V}$$
In the light
$$V_{out} = \{200/(200 + 1000)\} \times 6 \text{ V}$$
$$= 1 \text{ V}$$

The output voltage is V_{XY} and since Y is at zero potential the values calculated will also be *the potential at Y, V_Y.*

V_{out} can be made variable by making R_1 and R_2 one continuous resistor with a sliding junction point (the wiper) used for tapping off the output potential.

The circuit when used like this is called a *potentiometer.*

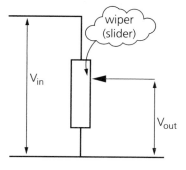

Example

Find the 3 currents and the 2 potential differences in the circuit below:

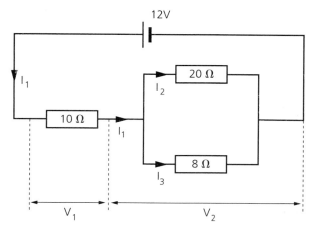

Stage 1
Reduce the parallel combination to the single equivalent resistor:

$$1/R_{tot} = 1/20 \ \Omega + 1/8 \ \Omega$$
$$1/R_{tot} = (7/40) \ \Omega^{-1}$$
$$R_{tot} = 5.7 \ \Omega$$

Stage 2
Re-draw the circuit using this value

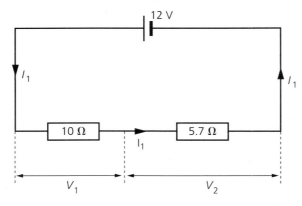

Stage 3
There are now 2 possible routes through – both equally valid:
either find the total current *first* using the complete circuit and *then* find the p.d.s
or regard the 2 series resistors as a *potential divider* and find the p.d.s *first*.

EITHER

$$I_1 = 12 \ \text{V}/(10 + 5.7) \ \Omega$$
$$= 0.76 \ \text{A}$$
$$V_1 = I_1 \times 10 \ \Omega = 7.6 \ \text{V}$$
$$V_2 = (12 - 7.6) \ \text{V} = 4.4 \ \text{V}$$
$$(\text{or } V_2 = I_1 \times 5.7 \ \Omega = 4.3 \ \text{V})$$

OR

> as a potential divider
> $$V_1 = \{10/(10 + 5.7)\} \times 12 \ \text{V}$$
> $$= 7.6 \ \text{V}$$
> $$V_2 = (12 - 7.6) \ \text{V} = 4.4 \ \text{V}$$
> (or, but more complicated,
> apply potential divider
> again to 5.7 Ω)

If p.d.s only are needed then the potential divider approach is recommended.

Stage 4
Bring the 2 approaches together again to find the remaining currents:

$$I_2 = V_2/20 \ \Omega = 0.22 \ \text{A}$$
$$I_3 = V_2/8 \ \Omega = 0.55 \ \text{A}$$
$$(\text{or } I_3 = I_1 - I_2)$$
Since the **or** method has
not yet found I_1
it can be found at this stage
by adding I_2 and I_3

Note that small 'rounding errors' lead to discrepancies in the 2nd sig.fig.

Kirchhoff's Laws

1st Law – currents at a junction
- No charge can build up at a junction in a circuit
- So in a given time interval all the charge that flows in to the junction flows out

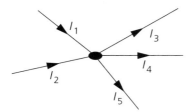

$I_1 + I_2 = I_3 + I_4 + I_5$
$I_1 + I_2 - I_3 - I_4 - I_5 = 0$
If the current flowing out is called a negative current
the Law is usually stated:

> total current into a junction is zero
> $\Sigma I = 0$
> (Σ means 'total')

2nd Law – loops of current
- The total change in potential in moving round a loop must be zero (conservation of energy).
- In moving in the *same* direction as (conventional) current through a resistor the potential *drops*.
- In moving through a cell *from* the –ve terminal *to* the +ve the potential *rises*.

The Law is usually stated:

> In moving round a loop in a circuit
> in the direction of conventional current
> the sum of the *IR* products for each resistor
> is equal to the *net* emf in the loop
> $\Sigma IR = \mathcal{E}$

Example
Find the currents in the two parts of the loop in the following circuit

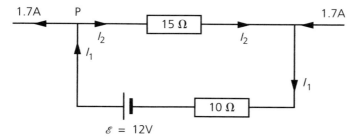

Stage 1

Form each '*IR*' product and add them.

Note that the units here are *volts* but when the algebra starts to get a little complicated it is permissible to drop them from every term as it is easy to lose the way otherwise! I_1 and I_2 now represent *pure numbers*, not a *value + unit* (see stage 6).

$$(I_1 \times 10) + (I_2 \times 15)$$

Stage 2

Note that the emf is +12 V.

(If the cell were connected the other way it would be −12 V.)

Stage 3

By K's 2nd Law:

$$10I_1 + 15I_2 = 12$$

Stage 4

Apply K's 1st Law to junction P:

$$I_1 = I_2 + 1.7$$

Stage 5

Substitute the expression for I_1 into the 2nd Law equation and solve for I_2:

$$10(I_2 + 1.7) + 15I_2 = 12$$
$$25I_2 = 12 - 17$$
$$I_2 = -0.2$$
$$\text{and hence } I_1 = +1.5$$

Stage 6

Final concluding statement with the proper units and reference back to the circuit:

current through the 10 Ω resistor is 1.5 A
current through the 15 Ω resistor is 0.2 A
in the *opposite direction* to that drawn (− sign)

4 Oscillations

The full difference is called 'peak-to-peak' - don't forget to halve it for the amplitude.

Terms used to describe oscillations:

- Cycle — the sequence of values of a quantity which forms the basic repetitive unit.
- Frequency f — number of complete cycles per second (unit: Hz or s^{-1}).
- Period T — the time for 1 cycle.
- Amplitude A — half the difference between the maximum and minimum value of the quantity.

Any quantity (e.g. velocity or acceleration) can have an amplitude, not just displacement.

$T = 1/f$

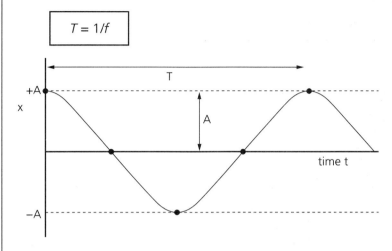

- For *mechanical* oscillations, x is the displacement of a particle from a fixed point.
- The most important type of oscillation is *Simple Harmonic Motion* (*SHM*) where T is *independent of A*.

SHM is described in terms of a *phase angle* θ.

- Over 1 cycle θ changes from 0 to 2π radians (360°).
- θ varies uniformly with time at a rate ω (radians / second, rad s^{-1} or just s^{-1}).
- After a time T (the period), θ has reached 2π rad so:

Important values of θ: at T/2 $\theta=\pi$ at T/4 $\theta=\pi/2$.

$$\omega = 2\pi/T$$
$$\omega = 2\pi f$$

- After a time t from the start of the motion, $\theta = \omega t$.

Quantitative properties of SHM

Displacement

- x varies *sinusoidally* with θ and therefore also with time.
- Whether it is the sine or cosine which is involved depends on the starting conditions (the graphs of the 2 functions have exactly the same shape).
- If the particle is displaced to value A and then released from rest:

'ωt' is the angle θ it has to be found first before taking the cosine.

$$x = A\cos\omega t$$

Example

A particle which moves with SHM is released from rest at a distance of 5.0 cm from the centre of the motion. After 0.4 seconds it has reached 1.2 cm from the centre.
What is (a) its amplitude (b) its frequency?

Stage 1

At $t = 0$, $\omega t (=\theta) = 0$ and $\cos 0$ is 1
$x = A\cos\omega t$
$5.0 \text{ cm} = A \times 1$
amplitude = 5.0 cm

Could state this directly since 5.0 cm is the maximum displacement.

Stage 2
Find ω first:

$1.2 \text{ cm} = 5.0 \text{ cm} \times \cos(\omega \times 0.4 \text{ s})$
$\cos(\omega \times 0.4 \text{ s}) = 0.24$
$\omega \times 0.4 \text{ s} = \cos^{-1}(0.24)$
$\omega \times 0.4 \text{ s} = 1.33 \text{ rad}$
$\omega = 3.3 \text{ rad s}^{-1}$

Make sure you know the sin and cos of 0 and 90°.

Calculator in radian mode! (don't forget to re-set it).

Stage 3
Find f:

$\omega = 2\pi f$
$f = \omega/2\pi$
$= 3.3 \text{ s}^{-1}/2\pi$
$= 0.53 \text{ Hz}$

A radian is a ratio without an SI unit – it doesn't feature in the final unit.

Velocity

The velocity is found from the gradient of the displacement–time graph. Its relationship to displacement is shown.

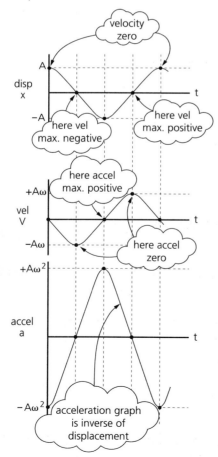

$$v_{max} = A\omega$$

When sketching these graphs always first locate the key points of zero, max. and min.

Acceleration

The acceleration is found similarly from the gradient of the velocity–time graph.
Relationship of acceleration graph to the displacement graph:
- has the same general shape
- but inverted (sign change)
- scaled by a factor ω^2

$$a = -\omega^2 x$$

Forces in SHM

The acceleration is important because by Newton's Second Law if you know acceleration you can find the force directly.

Example
If the mass of the particle in the above example is 0.3 kg, what force is acting after 0.4 s?

Stage 1
Calculation of acceleration:

> displacement x is 1.2 cm (0.012 m)
> $\omega = 3.3$ rad s^{-1}
> $\omega^2 = 10.9$ rad^2 s^{-2}
> $a = -\omega^2 x$
> $a = -10.9$ s$^{-2} \times 0.012$ m
> $= -0.13$ m s^{-2}
> (the sign indicates the acceleration is *towards* the centre)

Stage 2
Calculate the force:

> $F = ma$
> $= 0.3$ kg $\times (-0.13$ m s$^{-2})$
> $= -0.039$ N
> (all data, and hence the result, to 2 sig. figs.)

Factors affecting frequency

The frequency depends on:
- the *stiffness* of the system (force acting per unit displacement), k
- the mass m.

Sinc $F \propto a$ and $a \propto x$ then F must be proportional to x for the motion to be SHM, i.e. the stiffness must be constant (Hooke's Law)

- Increasing k increases f (larger force for the same displacement).
- Increasing m reduces f (larger inertia).

$$\omega = \sqrt{(k/m)} \text{ or } f = (1/2\pi)\sqrt{(k/m)}$$

Example
An empty lorry of mass 2000 kg bounces on its springs with a frequency of 1.2 Hz. It is filled with a load of sand of mass 5000 kg. What is the new frequency?

Method 1
Stage 1
Find the value of k:

$$f = (1/2\pi)\sqrt{(k/m)}$$
square both sides
$$f^2 = (1/4\pi^2)(k/m)$$
Multiply both sides by $4\pi^2 m$
$$4\pi^2 m f^2 = k$$
substituting the values gives
$$k = 4\pi^2 \times 2000 \text{ kg} \times (1.2 \text{ s}^{-1})^2$$
$$= 1.1 \times 10^5 \text{ N m}^{-1}$$

Calculator tip!
k is quoted to 2 sig. figs. as the data, but the 'full value' of k is left on display for calculation of stage 2: this reduces rounding errors.

Stage 2
Use this value to find the new frequency:

$$f = (1/2\pi)\sqrt{(1.1 \times 10^5 \text{ N m}^{-1}/7000 \text{ kg})}$$
$$= 0.64 \text{ Hz}$$
(Rough check: a lower value than the initial one is expected)

Method 2

$$\text{Since } f \propto 1/\sqrt{m}$$
$$f\sqrt{m} \text{ is a constant}$$
$$f_{new} \times \sqrt{(7000 \text{ kg})} = 1.2 \text{ Hz} \times \sqrt{(2000 \text{ kg})}$$
$$f_{new} = 1.2 \text{ Hz} \times \sqrt{(2000/7000)}$$
$$= 0.64 \text{ Hz}$$

If you are confident in handling proportionalities you could enter the calculation at this stage.

Energy changes in SHM

- During a quarter cycle from $x = 0$ to $x = A$ there is a steady transfer from kinetic to potential energy.
- During the next quarter cycle the change is reversed:
  At $x = 0$

E_k is a maximum and E_p is a minimum (usually taken as zero)
$$E_{k,max} = 1/2 \, mv_{max}^2$$
$$= 1/2 \, m(A\omega)^2$$
$$= 1/2 \, mA^2\omega^2$$

At $x = A$

E_k is zero and E_p is a maximum
by the Law of Conservation of Energy
E_p at the extremity must equal E_k at the centre
$$E_{p,max} = 1/2 \, mA^2\omega^2$$
$$= 1/2 \, mA^2(k/m)$$
$$= 1/2 \, kA^2$$

$$E_{tot} = 1/2 \, kA^2 \text{ or } 1/2 \, mA^2\omega^2$$

This is a very important result that total energy $\propto$ (amplitude)^2.

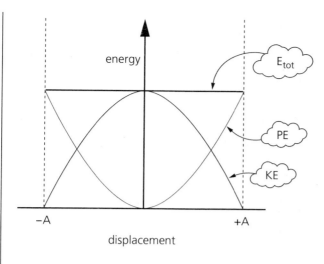

Energy changes during SHM

Example

A hydrogen atom has a mass of 1.7×10^{-27} kg and when bonded to an atom of chlorine in hydrogen chloride oscillates with a frequency of 7.7×10^{13} Hz. The oscillations are simple harmonic. The total energy is 2.0×10^{-20} J. What is the amplitude?

Method 1

Stage 1
Find k first:

$$f = (1/2\pi)\sqrt{(k/m)}$$
$$k = 4\pi^2 f^2 m$$
$$= 4\pi^2 \times (7.7 \times 10^{13}\ \text{Hz})^2 \times 1.7 \times 10^{-27}\ \text{kg}$$
$$= 4.0 \times 10^2\ \text{N m}^{-1}$$

Stage 2
Find A from the total energy:

$$E_{tot} = 1/2\ kA^2$$
$$A^2 = 2\ E_{tot}/k$$
$$A = \sqrt{(2\ E_{tot}/k)}$$
$$= \sqrt{(2 \times 2.0 \times 10^{-20}\ \text{J}/4.0 \times 10^2\ \text{N m}^{-1})}$$
$$= 1.0 \times 10^{-11}\ \text{m}$$

Method 2

Stage 1
Find ω first:

$$\omega = 2\pi f$$
$$= 2\pi \times 7.7 \times 10^{13}\ \text{Hz}$$
$$= 4.8 \times 10^{14}\ \text{s}^{-1}$$

Stage 2
Use total energy with this ω value to find A

$$E_{tot} = 1/2\ mA^2\omega^2$$
$$A^2 = 2\ E_{tot}/m\omega^2$$
$$A = \sqrt{(2\ E_{tot}/m\omega^2)}$$
$$= \sqrt{\{(2 \times 2.0 \times 10^{-20}\ \text{J}/(1.7 \times 10^{-27}\ \text{kg} \times 2.3 \times 10^{29}\ \text{s}^{-2})\}}$$
$$= 1.0 \times 10^{-11}\ \text{m}$$

Algebra point: since ω^2 is inside the $\sqrt{}$ it could come out at an earlier stage.

Free vibrations and damping

A system (which can be *modelled* as a mass on a spring) vibrating on its own with SHM is said to be undergoing *free vibrations* with a *natural frequency*

$f_n(= 1/2\pi\sqrt{\{k/m\}})$.

If there are no *dissipative* forces (friction or viscous drag) which generate a temperature rise the motion is *undamped* and will continue indefinitely at fixed amplitude.
The effect of such forces is to cause *damping* which:

- causes the amplitude to die away
- produces a *small* reduction in natural frequency which becomes more marked the heavier the damping

and, if the damping is *very* heavy

- removes the oscillation completely so that the initial displacement just decays to zero without any overshoot of the origin.

The transition between an oscillatory movement and one which just decays is called *critical damping*. A car's suspension is adjusted by the shock 'absorbers' (really dampers) to be around critical damping.

Forced vibrations and resonance

If a system capable of undergoing free vibrations is driven by some *external* force (the driver) at a driving frequency f_d, the motion of the system shows the following features.

- It vibrates with the *driving frequency* f_d (*not* its own natural frequency f_n).

The amplitude:
- depends on the amount of damping
- depends on how far away f_d is from f_n – when $f_d \gg$ or $\ll f_n$ the amplitude is small
- rises to a maximum (as f_d is varied) when $f_d = f_n$.

> This condition
> $f_d = f_n$
> is called *resonance*

The shape of the amplitude–driving frequency graph depends on the level of damping.

For low damping the graph:
- has a narrow peak (very low amplitude away from resonance)
- reaches a high (and potentially dangerous) maximum at resonance.

For high damping the graph:
- is much broader and shows a much less prominent peak
- shows a gradual reduction in frequency at the peak as damping is increased.

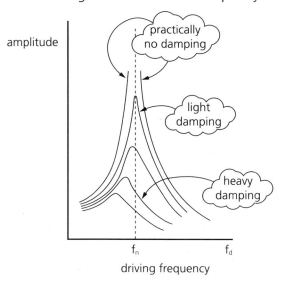

There may be a 'settling down' time when it vibrates with both frequencies showing a rising and falling amplitude (a 'beat').

Strictly speaking the condition that $f_d = f_n$ for resonance is not quite correct since f_n is the *undamped* natural frequency and, as the graph shows, the frequency at resonance reduces with increased damping: this is not normally anything to worry about since practical applications of resonance, when disaster can strike, involve very small levels of damping.

Examples of resonance

It can be a nuisance or potentially damaging in the following situations.

- Wind can induce forced vibrations onto structures such as bridges, tall buildings, cooling towers: the wind-induced frequency might coincide with a natural frequency (structures have more than one) causing resonance and possible structural failure.
- Loudspeakers have a natural frequency: if the damping were low they would amplify sounds preferentially in a narrow band around this frequency, producing distortion.
- Rotating machinery which is slightly out of balance can produce oscillatory forces which set up forced vibrations in the machinery itself or in nearby structures, again with the possibility of resonance and damage.

Resonance is important or essential:

- in the operation of musical wind instruments (see *standing waves* Chap. 5)
- when atomic bonding is investigated by using electromagnetic waves to excite resonance in vibrating molecules
- in magnetic resonance imaging (MRI) in medical applications
- in all radio tuning circuits, which rely on it (although this is electrical resonance – see Chap. 7 – not mechanical).

5 Waves

- A wave is a mechanism for the transmission of energy by a *linked* set of oscillators – one vibration is handed on to the next.
- The medium supporting a mechanical or electromagnetic wave does not itself move bodily – the wave passes through it.
- All the oscillators have the same
  – frequency
  – amplitude (in an undamped wave).

But in the direction in which the wave travels:

- there is a steady change in phase angle.

Phase difference and wavelength

2 points separated from each other along the distance of travel will have a *phase difference*.

- The phase difference between 2 points is the difference in their phase angles.
- Points which have a phase difference (in radians) of 0, 2π, 4π, ... $2n\pi$ (in degrees 0, 360, 720, ... $360n$) where n is an integer (0, 1, 2, 3, ...) are said to be *in phase*: at these points the oscillations are identical.
- The shortest distance (measured along the direction of travel) between 2 points in phase is called the *wavelength*, λ.
- Points separated by $\lambda/2$ are said to be in *anti-phase* – they have a phase difference of π, 3π, ... $(2n+1)\pi$.

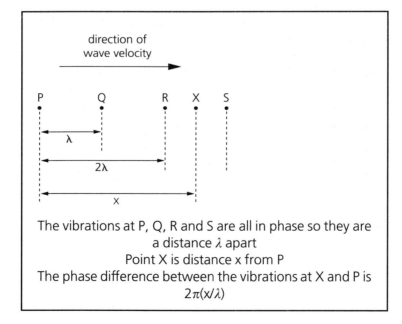

direction of
wave velocity

The vibrations at P, Q, R and S are all in phase so they are
a distance λ apart
Point X is distance x from P
The phase difference between the vibrations at X and P is
$2\pi(x/\lambda)$

Longitudinal and transverse waves

Longitudinal (also called *compression*) waves

- Vibrations of medium are parallel to direction of wave travel.
- Examples are sound and the P-seismic wave in earthquakes.

Revise the main parts of the electromagnetic spectrum and their position in it according to wavelength.

Transverse waves

- Vibrations of medium are perpendicular to the direction of wave travel.
- Examples are ripples on a liquid surface, waves on taut strings, S-seismic waves and all electromagnetic waves.

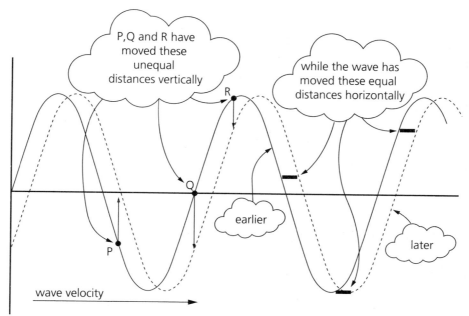

Progression of a transverse wave

Caution! graphs of y (displacement) against t (fixed x) and y against x (fixed t) have the same general shape – make sure you understand the difference: don't confuse period with wavelength.

Wavefronts and wave velocity

A wavefront is a *surface of constant phase* whose shape shows how the wave spreads out (propagates) from a source.

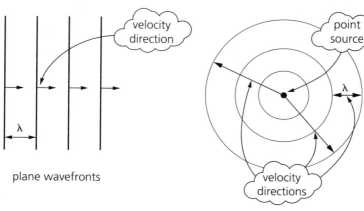

plane wavefronts

circular wavefronts

- From a point source they may be *circular* (in 2-dimensions, as with ripples) or *spherical* (in 3–dimensions).
- At a long distance from a point source (or from an extended flat source) the wavefronts will be *linear* (2-d) or *plane* (3-d).
- They are usually drawn one wavelength apart.
- The velocity *v* of a wave is the velocity at which a given wavefront moves and is always perpendicular to the wavefront:

$$\boxed{\begin{array}{l} \text{velocity} = \text{frequency} \times \text{wavelength} \\ v = f\lambda \end{array}}$$

Example

A sound wave in air has a velocity of 330 m s⁻¹ and transmits a musical note of frequency 440 Hz. What is the phase difference between 2 points separated along the direction of sound travel by (a) 0.50 m, (b) 1.75 m?

Stage 1
Calculate the wavelength:

$$\boxed{\begin{array}{c} v = f\lambda \\ \lambda = v/f \\ = 330 \text{ m s}^{-1}/440 \text{ Hz} \\ = 0.75 \text{ m} \end{array}}$$

Stage 2

$$\boxed{\begin{array}{c} \text{If the separation is } x \\ \text{phase difference } \delta = 2\pi x/\lambda \\ \text{(a) } \delta_1 = 2\pi \times 0.5 \text{ m}/0.75 \text{ m} = 4.2 \text{ radians} \\ \\ \text{(b) Since the separation is } > \lambda \text{ we expect a phase difference } > 2\pi \\ \delta_2 = 2\pi \times 1.75 \text{ m}/0.75 \text{ m} = 14.7 \text{ radians} \\ \\ \text{Since } 14.7 = 2\pi + 8.4 \\ \text{it would be correct to regard the phase difference as 8.4 radians} \\ \text{since } \cos 14.7 = \cos 8.4 \end{array}}$$

Refraction

The velocity of a wave depends on the medium through which it is travelling.
When a wave crosses a boundary between 2 media in which the velocities are different:
- the frequency stays the same
- the wavelength is changed (smaller in the slower medium)
- there is generally a direction change (unless the wave is travelling perpendicular to the boundary).

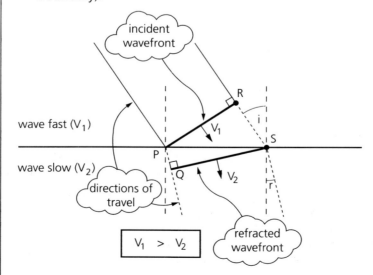

The wave travels from P to Q in the slow medium in the same time as it travels from R to S in the fast medium:

- i is the angle of incidence
- r is the angle of refraction
- the refractive index for this pair of media for the wave travelling *into* the slower medium is:

$$\mu = \sin i / \sin r = v_1 / v_2 \, (v_1 > v_2)$$

Refractive index varies with frequency, giving rise to the phenomenon of *dispersion*. In the visible spectrum, $\mu_{blue} > \mu_{red}$.

Total internal reflection

When a wave hits a boundary there is generally:
- some transmission through (the refracted wave)
- some reflection back from the boundary (the internally reflected wave).

How the incident power splits between these two waves depends on:
- the nature of the 2 media
- the direction of the incident wave.

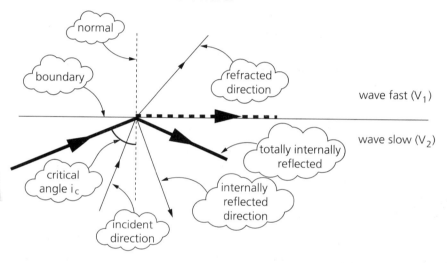

The transmitted power can be *zero* (all is reflected back) if:
- the incident wave is in the medium of *lower* velocity (v_2) **and**
- the angle of incidence is greater than a value called the *critical angle*, i_c.

> This condition is called *total internal reflection*
>
> $$\sin i_c = v_2 / v_1 = 1/\mu$$

Superposition

This term describes what happens when waves meet or cross over each other.

> **Principle of superposition**
> When two or more waves arrive at the same point
> the resultant disturbance is the vector
> sum of the disturbances
> which each wave would produce in the absence of the others

The simplest case is **2-source interference**.

The amplitude, A_{tot} of the disturbance at the point P where the 2 waves arrive depends on:

- the individual amplitudes A_1 and A_2
- the phase difference δ between the waves at P.

> If δ is $2\pi n$ radians
> the interference is *constructive* and
> $$A_{tot} = A_1 + A_2$$
> If δ is $(2n + 1)\pi$ (an odd multiple of π)
> the interference is *destructive* and
> $$A_{tot} = A_1 - A_2$$

As a special case, if $A_1 = A_2 = A$ then:

- for constructive interference $A_{tot} = 2A$
- for destructive interference $A_{tot} = 0$.

The simple experimental arrangement

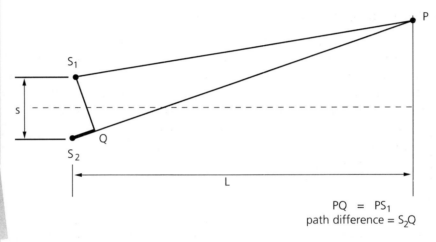

$$PQ = PS_1$$
path difference = S_2Q

This is an important diagram which appears often in discussion of interference – you need to understand path diff. and how it produces phase difference.

- S_1 and S_2 are 2 point sources of waves (of *any* kind).
- They are in phase.
- P is a point where the 2 sets of waves overlap.
- The phase difference at P depends on the *path difference*, $S_2P - S_1P$.
- The path difference is shown as S_2Q.

Path difference and phase difference

Provided S_1 and S_2 are in phase, the interference at P is:

- constructive if $S_2Q = n\lambda$ (whole number of wavelengths)
- destructive if $S_2Q = (2n + 1)\lambda/2$ (odd number of half wavelengths).

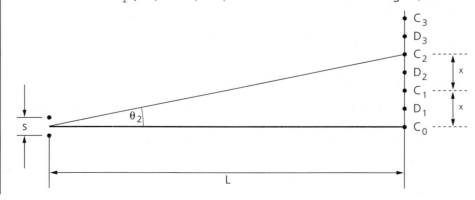

As the point of observation P is moved parallel to S_1S_2 it passes through successive places of constructive (C_0, C_1, C_2, . . .) and destructive interference (D_1, D_2, . . .):

- for sound waves these are successive points where the *loudness* is a maximum or minimum (or zero)
- for light they are successive points of bright and dark usually called *interference fringes*.

You need to know the factors which determine x and how they affect it – check with your exam syllabus.

> Provided $s \ll L$ the maxima and minima are equally spaced
> and the *fringe spacing x*
> (one max. to the next or one min. to the next)
> is given by
> $x = \lambda L/s$

Rather than work in distance x it is sometimes more convenient to work with *direction*, the angle θ as shown.

> If θ_n is the angle for the nth point of constructive interference C_n
> $\sin \theta_n = n\lambda/s$

The phases of the sources – coherence

It is *not essential* that S_1 and S_2 are in phase. If they are in antiphase:
- the points of constructive and destructive interference change round
- the spacing of the pattern remains the same.

It *is essential* though that S_1 and S_2 have a *constant* phase relation in order to observe a steady interference pattern, but this phase relation (difference) could have any value.

> When 2 sources have a constant phase relation
> they are said to be *coherent*
> (otherwise *incoherent*)

- Steady sound sources like loudspeakers will usually be coherent.
- Light sources can only be coherent if the two sets of waves originated from the *same wavefront* (e.g. one wave passing through 2 narrow slits and dividing the wavefront).

Intensity

The intensity of a wave is the power transmitted per unit area (unit: W m^{-2}) perpendicular to the direction of propagation.
- For sound, intensity is related to loudness.
- For light, intensity is related to brightness.
- From a point source the *intensity* $\propto 1/(distance)^2$

Just as for SHM where the *total energy* $\propto$ *(amplitude)^2* so for a wave,

intensity $\propto$ (amplitude)^2

Therefore at a place of constructive interference of 2 waves of equal amplitude the intensity is 4 times greater than that of one of the waves by itself.

Standing (or stationary) waves

These result from the interference of 2 sets of waves of the same frequency passing through each other in *opposite directions*.
- They will occur when waves are confined by boundaries in a fixed region (length, area or volume).
- The interference is between the waves reflected from the boundary and the incident waves.

- The resulting wave pattern remains *fixed in space* (hence stationary).
- Through the wave the amplitude varies continuously from *zero* (destructive interference, the *displacement nodes*) to a *maximum* (constructive interference, the *displacement antinodes*).
- The nodes and antinodes are normally equally spaced – one node to the next is $\lambda/2$.

Standing waves on a taut string

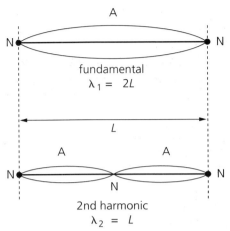

- The 2 ends are *always* displacement nodes.
- The longest wavelength (lowest frequency) occurs when there is just one antinode in the middle.
- The pattern of this vibration is called the *fundamental mode* and the frequency the fundamental frequency, f_1.
- $L = \lambda_1/2$ and so the fundamental wavelength $\lambda_1 = 2L$
- Other modes are possible (called *harmonics*) by fitting more nodes in – the 2nd harmonic has a node in the middle.
- There is an infinite sequence of harmonics with frequencies whole number multiples of the fundamental.

Example

A wire of length 1.2 m is made to vibrate in a standing wave with 3 loops on it at a frequency of 420 Hz. What is the velocity of the transverse wave on the string?

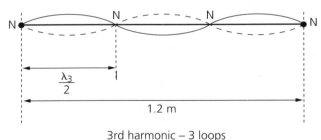

3rd harmonic – 3 loops

Stage 1
Draw a diagram and find the wavelength:

> The string is vibrating at its 3rd harmonic so call
> the frequency and wavelength f_3 and λ_3
> $\lambda_3/2 = (1/3) \times 1.2$ m (i.e. length of 1 loop)
> $\lambda_3 = 0.80$ m

Stage 2
Use $v = f\lambda$:

$$v = f_3\lambda_3$$
$$= 420\text{ Hz} \times 0.80\text{ m}$$
$$= 336\text{ m s}^{-1}$$

So velocity is 340 m s^{-1} to the appropriate accuracy

Standing waves in an air-filled pipe

- An open end is always a displacement antinode.
- A closed (stopped) end is always a displacement node – air can't move.
- The possible standing wave patterns depend on whether there are 1 or 2 open ends.
- The *principles* for finding the harmonics are the same as those for a taut string.

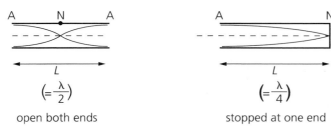

$$\left(=\frac{\lambda}{2}\right) \qquad \left(=\frac{\lambda}{4}\right)$$

open both ends   stopped at one end

An open and a stopped pipe showing the amplitude variation along the pipe in the fundamental mode. The distance apart of the 2 curved lines at any point represents the amplitude at that point.

Caution! It is easy to be confused by a diagram such as this – don't forget that the actual vibrations are *along* the axis of the pipe.

Example
A pipe of length 0.40 m is stopped at one end. What are the first 2 frequencies at which standing waves are set up inside? Speed of sound = 330 m s^{-1}

Stage 1
Find the longest wavelength λ_1

$$\lambda_1/4 = L$$
$$\lambda_1 = 4 \times 0.40\text{ m} = 1.6\text{ m}$$

Stage 2
Find the next wavelength λ_2:

The length of the pipe (A–N–A–N) is
$$\lambda_2/4 + \lambda_2/2 = 3\lambda_2/4$$
$$3\lambda_2/4 = 0.40\text{ m}$$
$$\lambda_2 = 0.53\text{ m}$$

A refinement to the story or a Grade A! you need to be aware whether the wave is being described by displacement or pressure: an antinode for one is a node for the other – some microphones respond to pressure, some to displacement.

Refer back to your textbook if you are unsure about the displacement and pressure fluctuations in a sound wave.

Stage 3
Use $v = f\lambda$ to find both frequencies:

$$f_1 = 330 \text{ m s}^{-1}/1.6 \text{ m}$$
$$= 206.25 \text{ Hz}$$
$$f_2 = 330 \text{ m s}^{-1}/0.53 \text{ m}$$
$$= 622.64 \text{ Hz}$$

Since the data is to 2 significant figures the frequencies
should be rounded to this accuracy
i.e. 210 Hz and 620 Hz
(N.B. f_2 is actually $3f_1$ but the rounding to 2 sig. figs.
at each stage does not show this accurately)

Diffraction

This effect is a property of *all* waves. It is the change in shape of a wavefront when the wave passes through gaps or around obstacles. It results in the *spreading out* of a wave – if there is much diffraction the waves cease to travel in straight lines.

The amount of spreading (measured by the angle of the diverging beam) depends on:
- the wavelength λ
- the aperture width b (or obstacle size).

The amount of diffraction is determined by
the ratio λ/b

If different combinations of wavelength and aperture
have the same value for this ratio then
the waves will diffract identically

- Waves with *large* wavelengths diffract *more* than those with shorter.
- Smaller apertures produce *greater* diffraction than large ones.
- Diffraction ultimately limits the ability to see fine detail (*resolution*).
- Short wavelengths can therefore be used for greater resolution.
- As a rough guide, if λ is comparable in size to b then diffraction is likely to be important.

Example
The ability of an optical telescope to resolve fine astronomical detail is determined by the wavelength λ of the radiation used and the diameter b of the telescope lens or mirror. A typical combination is $\lambda = 5.0 \times 10^{-7}$ m and $b = 2.0$ m.
A radio telescope is used to detect radio waves of wavelength 0.2 m. What diameter should it have so that diffraction effects are the same as for the optical telescope (i.e. so that its resolution is the same)? Comment on the value.

Stage 1
Calculate the determining factor λ/b:

$$\lambda/b = 5.0 \times 10^{-7} \text{ m}/2.0 \text{ m}$$
$$= 2.5 \times 10^{-7} \text{ (no units)}$$

You will not need to reproduce the detailed theory of a single slit but you need to be quite familiar with the factors and how they influence diffraction.

Stage 2
Compare with the radio telescope to find its value of *b*, say b_r:

$$0.20 \text{ m}/b_r = 2.5 \times 10^{-7}$$
$$b_r = 8.0 \times 10^5 \text{ m (800 km)}$$

Stage 3
Comment:

Clearly 800 km is impossible
A typical 'dish' may only have $b = 25$ m
It can only resolve $25/8 \times 10^5$ (=1/32000)
as well as the optical telescope

This is why small dishes are set up in arrays several km apart, producing the equivalent effect of one huge diameter dish.

Diffraction grating

- Consists of several thousand parallel slits (actually grooves in glass or transparent film) with a separation *s* (*not* the width of each slit).
- Relies on diffraction at each slit and then interference of light from adjacent slits.
- So the theory in principle is the same as for 2-source interference.
- The maxima are observed in the transmitted beam at angles θ_n given by:
  $$n\lambda = s \sin\theta_n$$
  (*n* is an integer called the order of the interference)

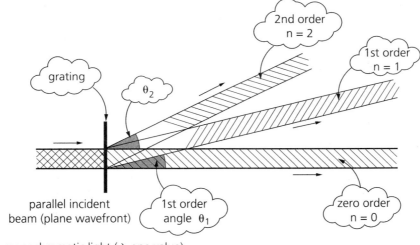

monochromatic light (λ one value)

Check if your syllabus requires you to be able to use a spectrometer with a grating to measure wavelength accurately.

Three important differences from the 2-source case are:
- the interference pattern is much *brighter*
- the maxima are much *sharper*, so are easier to locate
- the angles θ_n are much *larger* (increasing the accuracy of measurement) because *s* is much smaller.

These three improvements make the diffraction grating the standard method for accurate measurement of wavelength.

Polarisation

The term applies to *transverse waves only* (of all kinds).
The two directions
– velocity of wave
– direction of vibration of whatever is oscillating

define a *plane* (a flat surface) called the *plane of polarisation*.

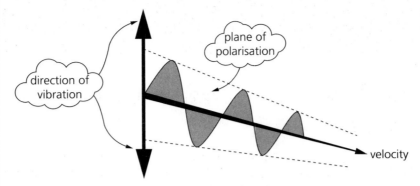

- If this plane remains fixed in space the wave is *plane polarised*.
- If the plane rotates at a steady rate the wave is *circularly* polarised.
- If the plane of polarisation changes randomly and rapidly so that it has no preferred direction the wave is *unpolarised*.

Light waves from usual light sources (including the Sun) are unpolarised. If they are passed through some natural or artificial ('polaroid') crystalline materials they can be made plane polarised.

Reflected unpolarised light is at least partially plane polarised (some randomness on top of a preferred direction).

Polarisation and scattering of light

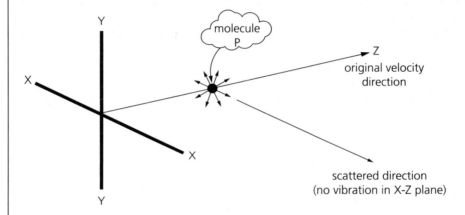

Light travelling in the Z-direction hits a molecule P, causing it to vibrate and re-radiate light. The light scattered in the X-direction (perpendicular to the original direction):
- cannot have any vibration in the X-direction (transverse)
- cannot have any vibration in the Z-direction (none in the original wave)
- so must be plane polarised in the Y-direction.

The blue light from the sky is produced in this way and is plane polarised if viewed at 90° to direct sunlight.

6 Gravitational and electric fields

Uniform fields

Gravitational field

- A gravitational field is produced *by a mass* (or a collection of masses).
- The field exerts a force *on a mass* and the field is in the same direction as the force.
- The field strength **g** (a vector) is the force acting *per unit mass*:

$$\textbf{g} = \textbf{F}/m \text{ (unit: N kg}^{-1}\text{)}$$
$$\text{this is often re-arranged to}$$
$$\textbf{F} = m\textbf{g}$$

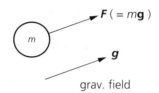

grav. field

If **g** is the gravitational field due to the Earth or some other astronomical object then **F** is called the *weight* **W**.

- The field is uniform if its magnitude *and* direction remain constant from one place to another.
- The field *close to the Earth's surface* is reasonably uniform,
  $g_{surf} \approx -9.8\text{ N kg}^{-1}$
  with the sign convention that positive is away from the Earth (upwards)

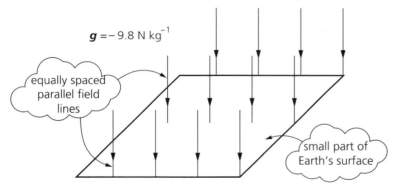

Uniform gravitational field near the Earth

If the gravitational field produces the *only force* acting on a mass (a situation described as *free fall*) and produces an acceleration **a**, then:

$$\textbf{F} = m\textbf{a}$$
$$\text{and by the definition of field strength}$$
$$\textbf{F} = m\textbf{g}$$
$$\text{so } \textbf{a} = \textbf{g}$$
$$\text{for free fall}$$

So the local free fall acceleration is:

- independent of the mass (all masses have the same acceleration)
- equal to the local value of the gravitational field strength (1 m s^{-2} = 1 N kg^{-1}).

Electric field (E-field)

- An E-field is produced *by a charge* (or a collection of charges).
- The field exerts a force *on a charge*.
- The field strength **E** (a vector) is the force acting per unit charge on a *positive charge, q*:

> **E** = **F**/q (unit: N C^{-1})
> this is often rearranged to
> **F** = **E**q

- The field is uniform if its magnitude *and* direction remain constant from one place to another.
- An E-field is uniform close to a *flat* charged conducting surface (except near the edges)
- It is shown pictorially as *field lines* which start on a +ve charge and end on a −ve charge.

Capacitors – a first look

A very common arrangement is to use 2 flat conducting plates each of area *A* parallel to each other and carrying equal and opposite charges, +*Q* and −*Q*. If they are a distance *d* apart:

> *E* = *V*/*d*

giving an alternative unit for *E* as *volt per metre* (V m^{-1})
This arrangement is a *capacitor*.

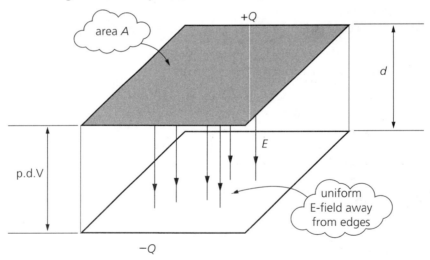

Experiments show that:
$Q \propto V$
$Q \propto A$
$Q \propto 1/d$
where V is the p.d. between the plates.

Combining these into one overall relation:

$Q \propto VA/d$

If the space between the plates is a vacuum the constant of proportionality is written ε_0 ('epsilon nought'):

$$Q = \varepsilon_0 VA/d$$

ε_0 is called the permittivity of free space and has the value
8.85×10^{-12} C V^{-1} m^{-1}

The ratio Q/V (= $\varepsilon_0 A/d$) is a constant for a given size and spacing of plates called the *capacitance C.*

For a parallel plate capacitor with a vacuum

$C = \varepsilon_0 A/d$ (unit: coulomb/volt or farad, F)

- If any other *insulating* material is between the plates the capacitance increases by a factor ε_r, the *relative permittivity* of the material.

The product $\varepsilon_0 \varepsilon_r$ is called *the permittivity* of the material.

Example

A small drop of oil of mass 4.0×10^{-15} kg picks up an electron of charge -1.6×10^{-19} C. The drop is in a vertically downwards E-field produced by putting 5 kV across two parallel plates 15 mm apart. If the only forces acting on the drop are gravity and the electric force on the electron, what is the acceleration of the drop? ($g = -9.8$ N kg^{-1})

Stage 1
Calculate the E-field strength:

$$E = V/d$$
$$= 5000 \text{ V}/15 \times 10^{-3} \text{ m}$$
$$= 3.3 \times 10^5 \text{ V m}^{-1} \text{ (or N C}^{-1}\text{)}$$

Since the field is downwards
$$\boldsymbol{E} = -3.3 \times 10^5 \text{ V m}^{-1}$$

Stage 2
Calculate the force on the electron:

$$\boldsymbol{F} = \boldsymbol{E}q$$
$$= -3.3 \times 10^5 \text{ N C}^{-1} \times (-1.6 \times 10^{-19} \text{ C})$$
$$= +5.3 \times 10^{-14} \text{ N}$$

Stage 3
Find the weight of the drop:

$$\boldsymbol{W} = m\boldsymbol{g}$$
$$= 4.0 \times 10^{-15} \text{ kg} \times (-9.8 \text{ N kg}^{-1})$$
$$= -3.9 \times 10^{-14} \text{ N}$$

Stage 4
Apply Newton's Second Law, where F is the resultant force of the gravitational and electric fields on the drop:

$$\boldsymbol{F} = m\boldsymbol{a}$$
$$+5.3 \times 10^{-14} \text{ N} + (-3.9 \times 10^{-14} \text{ N}) = 4.0 \times 10^{-15} \text{ kg} \times \boldsymbol{a}$$
$$\boldsymbol{a} = +1.4 \times 10^{-14} \text{ N}/4.0 \times 10^{-15} \text{ kg}$$
$$= +3.5 \text{ m s}^{-2}$$

The acceleration is *upwards* (+ sign)

farad is huge! (although it does exist) if you get an answer of a few farads check the power of 10.

Does your syllabus require knowledge of the Millikan oil drop experiment?

A downwards E-field exerts an upwards force on a negative charge.

Note the full use of signs here – used with confidence and practice you will always get the direction right.

Accelerating a charge

- If a charged particle of mass m and charge q is accelerated from rest through an accelerating voltage V then the gain in KE is equal to the work done by the electric field (or the *loss* in electrical P.E.), so that:

$$1/2\ mv^2 = qV$$

- If the charge has a component of velocity perpendicular to the field then its motion can be treated in exactly the same way as a projectile under gravity (Chap. 1) and it will follow a *parabolic* path.

Non-uniform fields

Radial gravitational field

- Is produced by a point mass or a spherical distribution of mass (the Earth, Moon, Sun etc.).
- The field direction is *towards* the centre of the mass.
- The field strength follows an *inverse square law* of distance from the centre.
- Cannot be shielded by any intervening medium.

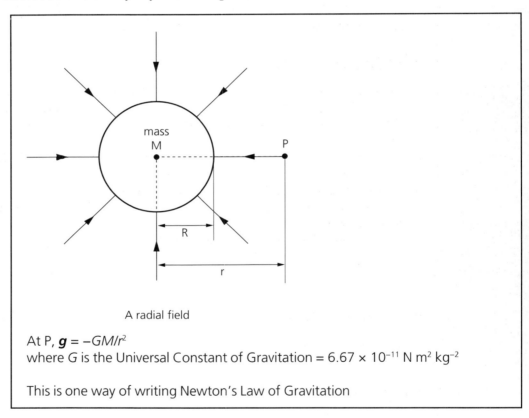

mass
M

P

R

r

A radial field

At P, $\boldsymbol{g} = -GM/r^2$
where G is the Universal Constant of Gravitation $= 6.67 \times 10^{-11}$ N m² kg⁻²

This is one way of writing Newton's Law of Gravitation

Severe health warning! this result only applies for $r \geq R$ i.e. not 'inside' – what would you expect g to be at the centre? –what does the equation give you if $r=0$?.

Example
The Moon's radius R_m is 1.740×10^6 m and has a surface gravitational strength of -1.62 N kg⁻¹. Use the data to find the Moon's mass M_m.

Since the data apply to the surface ($r = R_m$)
the expression for $\boldsymbol{g}$ in terms of r can be used directly
$$\boldsymbol{g}_m = -GM_m/R_m^2$$
$$M_m = -\boldsymbol{g}_m R_m^2/G$$
$$=-(-1.62 \text{ N kg}^{-1}) \times (1.740 \times 10^6 \text{ m})^2/6.67 \times 10^{-11} \text{ N m}^2 \text{ kg}^{-2}$$
$$= 7.35 \times 10^{22} \text{ kg}$$

Data to at least 3 sig. figs. so result .....

Force of attraction between point or spherical masses, M and m:

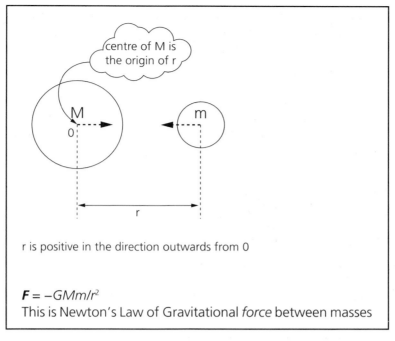

r is positive in the direction outwards from 0

$F = -GMm/r^2$
This is Newton's Law of Gravitational *force* between masses

The minus sign *always* accompanies an *attractive* force.

Application to satellites

You need to be clear how one step leads to the next

.

.

.

.

but don't try and remember this result – understand how it is made up.

- Same principle applies equally to artificial satellites or natural satellite systems – the Moon around Earth, other planetary moons, the individual planets around the Sun.
- An object moving in a circle of fixed radius r at constant *speed v* is accelerating (direction changes so velocity changes).
- The acceleration is constant in magnitude and is directed at any instant along the radius *towards* the centre (*centripetal* and is given a minus sign).
- The acceleration is $-v^2/r$ (or $-\omega^2 r$ where ω is the *angular* velocity).
- The only force on a satellite is gravity so it is in free fall with acceleration equal to the local value of **g**.
- The centripetal acceleration must therefore be the *same* as **g**, so:

$$-GM/r^2 = -v^2/r$$

or

- the same result is arrived at by applying Newton's Second Law to the satellite:

force = mass × acceleration
$-GMm/r^2 = m \times (-v^2/r)$

from which it is seen that the value of m disappears and is irrelevant.

Example
A satellite is launched very close to the Earth's surface (just outside the atmosphere). How fast is it travelling? Earth's radius R_E is 6400 km.

The local value of **g** is still very close to -9.8 m s^{-2}
$-v^2/R_E = -9.8$
$v^2 = 9.8 \text{ m s}^{-2} \times 6.4 \times 10^6 \text{ m}$
$= 6.3 \times 10^7 \text{ m}^2 \text{ s}^{-2}$
$v = 7.9 \times 10^3 \text{ m s}^{-1}$ (7.9 km s^{-1})

What does the speed become if the radius of the orbit is doubled to $2R_E$?

Method 1

Start with $-GM_E/r^2 = -v^2/r$
It looks as if the values of G and M_E are needed but in fact the product GM_E can be found from the Earth's surface data
$-GM_E/R_E^2 = -9.8$ m s^{-2}
giving $GM_E = 4.0 \times 10^{14}$ m^3 s^{-2}
Putting this value into the centripetal acceleration equation and multiplying through by r (which now $= 2R_E$) gives
$v^2 = 4.0 \times 10^{14}$ m^3 s$^{-2}/(2 \times 6.4 \times 10^6$ m$)$
$v = 5.6 \times 10^3$ m s^{-1}

Method 2

Apply the inverse square law directly
to find the value of $\boldsymbol{g}$ at a distance of twice the Earth's radius
r has doubled so the magnitude of $\boldsymbol{g}$ has reduced by $(1/2)^2$, i.e. $1/4$
New $\boldsymbol{g} = -9.8/4$ m s$^{-2} = -2.45$ m s^{-2}
so $-v^2/(2R_E) = -2.45$ m s^{-2}
$v = 5.6 \times 10^3$ m s^{-1}

Potential energy in a radial gravitational field

- A pair of masses has gravitational energy because work has to be done to separate them.
- The energy increases as they get further apart.
- The zero of PE is taken to be when they are at an infinite separation.
- So the energy at any other separation must be negative.

> The negative sign for PE is always an indicator of an attractive force.

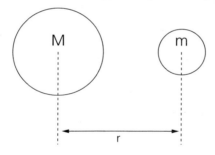

> Caution! only single power of r on the bottom line for energy.

$E_p = -GMm/r$
The potential energy per unit mass is called the gravitational potential V_g
$V_g = E_p/m = -GM/r$ (unit: J kg^{-1})

Escape velocity

An object will just escape from the gravitational field of a large mass such as the Earth if it is given enough KE which, when added to the (negative) PE, produces a total energy of zero. The velocity for this to happen is the escape velocity $\boldsymbol{v_e}$.
By the Law of Conservation of Energy this means that when it reaches infinity it will have zero PE (by definition) and must have also zero KE.

> This does not depend on m which cancels out from each side.

$E_{tot} = 1/2\, mv_e^2 + (-GMm/r) = 0$
from which $v_e = \sqrt{(2GM/r)}$

Example

What is the magnitude of the escape velocity from the Moon's surface?
$R_m = 1740$ km and $M_m = 7.35 \times 10^{22}$ kg

$$v_e = \sqrt{\{(2 \times 6.67 \times 10^{-11}\ \text{N m}^2\ \text{kg}^{-2} \times 7.35 \times 10^{22}\ \text{kg})/1740 \times 10^3\ \text{m}\}}$$
$$= \sqrt{(5.64 \times 10^6\ \text{kg m s}^{-2}\ \text{m}^2\ \text{kg}^{-2}\ \text{kg m}^{-1})}$$
$$= 2370\ \text{m s}^{-1}$$

Equipotentials

- These are surfaces joining points having the same gravitational potential – they can be thought of as contours of energy for a mass of 1 kg rather like map contour lines.
- There is no energy change in moving along an equipotential so the field always crosses them at 90°.
- In 2-dimensions the equipotentials of a radial field are concentric circles.

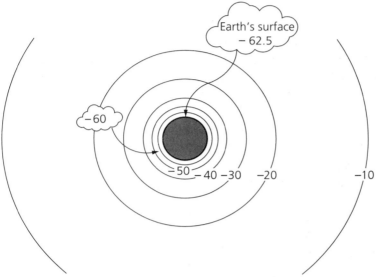

Equipotentials of the Earth/MJ kg^{-1}

Radial electric field

- Is the field produced by a *point charge* or a charged spherical conductor.
- The field direction depends on the sign of the charge producing it.

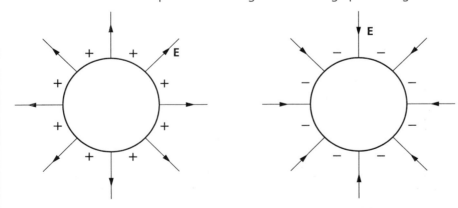

- The field strength follows an *inverse square law* (Coulomb's Law) of distance from the centre.

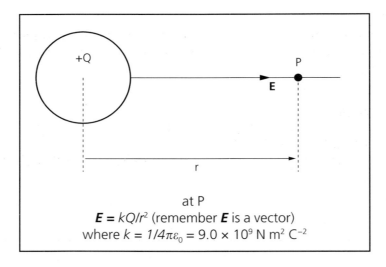

at P
$E = kQ/r^2$ (remember E is a vector)
where $k = 1/4\pi\varepsilon_0 = 9.0 \times 10^9$ N m^2 C^{-2}

If the sign of Q is used correctly the direction of the field automatically follows.

Force between charges

If a charge q is placed at P the force on it is:

$F = Eq$ (definition of field strength)
$F = (1/4\pi\varepsilon_0)Qq/r^2$

A positive force is repulsion either both + or both −.

Potential energy in a radial electric field

- A pair of charges has electrostatic potential energy because work is done to pull them apart (opposite signs) or push them together (same signs).
- The zero of potential energy is taken to be when they are at an infinite separation.
- So the energy at any other separation must be negative (attraction) or positive (repulsion).

Again note that the signs of Q and q give the correct sign for energy.

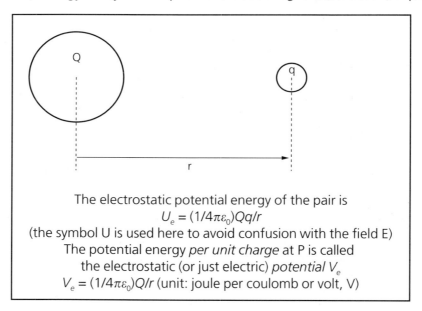

The electrostatic potential energy of the pair is
$U_e = (1/4\pi\varepsilon_0)Qq/r$
(the symbol U is used here to avoid confusion with the field E)
The potential energy *per unit charge* at P is called
the electrostatic (or just electric) *potential* V_e
$V_e = (1/4\pi\varepsilon_0)Q/r$ (unit: joule per coulomb or volt, V)

Caution (again!) single power of r on the bottom as for gravitation.

Equipotentials for a radial electric field are surfaces of equal electric potential and just like the gravitational case can be thought of as contours of equal electric potential energy of a 1 coulomb charge. They relate to the field in exactly the same way as gravitational equipotentials.

The electronvolt (eV) is a unit of energy commonly used when the charges are atomic particles (electrons, protons, ions):

1 eV is the change in energy when 1 electronic charge e (=1.6×10^{-19} C) is moved through a p.d. of 1 volt
1 eV = 1.6×10^{-19} J

1 C is an enormous charge - you cannot actually put it at a point (but neither do you actually need to travel for 1 hour to have a speed in miles per hour).

Example

How much energy is required to pull the electron in a hydrogen atom completely away from the proton (i.e. to ionise the atom) if they are initially 1.0×10^{-10} m apart? Give the value in electronvolt.

Stage 1

Interpret the wording of the question!
- 'Completely away' should be taken to mean 'remove to infinity'.
- So the final potential energy is zero.
- So the energy *change* in moving from $r = 1.0 \times 10^{-10}$ m must just be the value of the potential energy at this distance.

Stage 2

Calculation of energy, taking $Q = +1.6 \times 10^{-19}$ C and q as -1.6×10^{-19} C:

$$\text{potential energy} = (1/4\pi\varepsilon_0)Qq/r$$
$$= 9.0 \times 10^9 \text{ N m}^2 \text{ C}^{-2} \times (+1.6 \times 10^{-19} \text{ C}) \times (-1.6 \times 10^{-19} \text{ C})/1.0 \times 10^{-10}\text{m}$$
$$= -2.3 \times 10^{-18} \text{ N m}$$

Stage 3

Interpret the result and convert the units:
- 1 N m is 1 J so the units are those of energy
- the energy is negative, so to remove the electron to infinity (zero PE) requires the *addition* of 2.3×10^{-18} J
- using 1 eV = 1.6×10^{-19} J, the energy required (the *ionisation energy*) becomes 14.4 eV
- decide the number of significant figures – the data is to 2 so the result should be to 2, i.e.14 eV.

Comparison of gravitational and electric radial fields and potentials

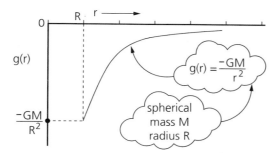

Gravitational field (Nkg^{-1} or ms^{-2})

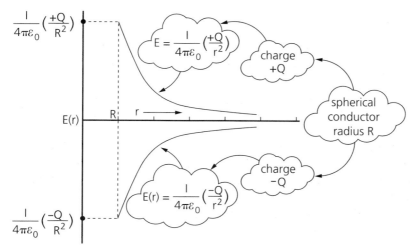

Electric field (NC^{-1}or Vm^{-1})

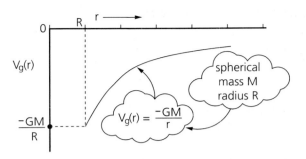

Gravitational potential (Jkg^{-1})

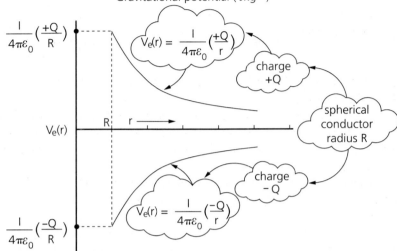

Electric potential (JC^{-1} or volt)

More on capacitors

Series and parallel combinations of C_1, C_2, C_3 . . .

- Series: $\quad\quad\quad 1/C_{tot} = 1/C_1 + 1/C_2 + 1/C_3 + . . .$

- Parallel: $\quad\quad C_{tot} = C_1 + C_2 + C_3 + . . .$

Check if your syllabus requires you to derive these.

Stored energy

If a capacitor is charged up to the values represented by point P on the V–Q graph it has stored energy which is the *area* under the graph up to P.

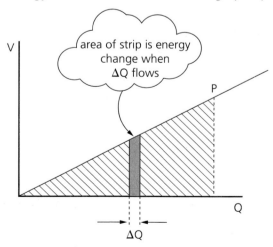

$$E_p = 1/2 \; QV$$
Using $C = Q/V$
the energy also becomes
$$E_p = 1/2 \; CV^2 \text{ or } 1/2 \; (Q^2/C)$$

Discharge of a capacitor through a resistor

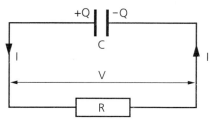

- V is the same for both R and C.
- So $IR = Q/C$.
- Since current is rate of flow of charge, I can be written as $-\Delta Q/\Delta t$ (Q getting smaller so a negative sign):

> this relation becomes
> $$\Delta Q/Q = -\Delta t/RC$$

- For equal time intervals the *proportional* changes in Q are equal – this is the fundamental property of an *exponential* change with time.
- The product RC has units of seconds and is called the *time constant* of the circuit.
- RC is a measure of how rapidly the discharge lasts – a useful rough rule is that it takes about 4 time constants to discharge (i.e. for Q or I to reach nearly zero – theoretically they *never* reach zero).
- After 1 time constant the charge (or current) has fallen to a fraction $1/e$ of its initial value – 0.37 or 37%.

You need to know the basic properties of exponential changes.

$e \approx 2.7$ so $1/e \approx 0.37$.

You are unlikely to be asked to do complicated algebra on exponentials in an exam but you do need to be really familiar with time constant – it is the quickest way of handling anything to do with time of charging or discharging.

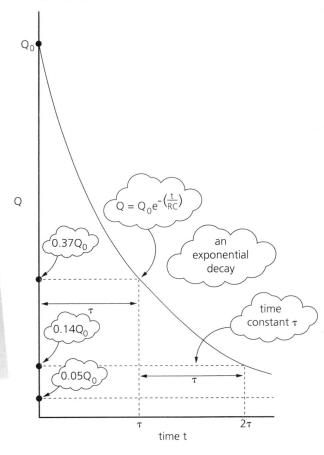

Discharge graph for a capacitor
$$Q = Q_0\, e^{-(t/RC)}$$

V and I follow the same pattern since they are both $\propto Q$.

If τ is the time constant, the table below shows how much charge is left after whole numbers of time constants:

time	0	τ	2τ	3τ	4τ
charge	Q_0	$0.37Q_0$	$0.14Q_0$	$0.05Q_0$	$0.02Q_0$

7   Magnetic effects and currents

Magnetic fields are *produced* by:
- permanent magnets
- a moving charge (often a current in a wire).

Magnetic fields can be *detected* by a freely pivoted compass needle (a *dipole*) put in the field.

The *direction* of a magnetic field at any place is the direction in which the *North-seeking* pole (often just called the North pole) of a compass points.

> The current does not have to be confined to a conductor – any moving charge will do.

Shapes of magnetic fields

- Magnetic fields can be represented by continuous *field lines*.
- The direction of the field line at any point gives the local direction of the field.
- Field lines never cross or touch.

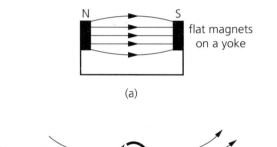

(a)

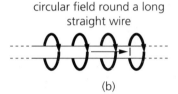

circular field round a long straight wire

(b)

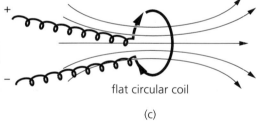

flat circular coil

(c)

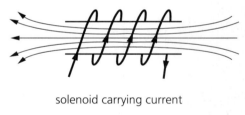

solenoid carrying current

(d)

> The N-seeking pole of a compass is that end which points roughly to geographical North when well away from any field other than the Earth's.

Arrangements (b), (c) and (d) are best drawn for clarity with the convention:

$\otimes$ away $\odot$ towards

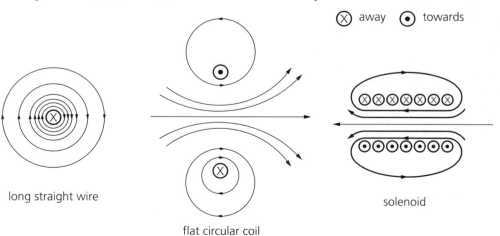

long straight wire

flat circular coil

solenoid

> Note that despite the apparent contradiction in (a) magnetic field lines are continuous loops.

The solenoid field (well inside it) is the best way of producing a *uniform magnetic field*.

Forces produced by magnetic fields

On a wire

- A magnetic field produces a force on a wire carrying a current in the field provided the wire is *not parallel* to the field.

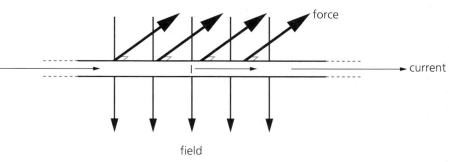

The force *F* is:
- always perpendicular to the wire
- proportional to the current, *I*
- proportional to the length of wire in the field, ℓ

So

$F/I\ell$
is constant for a given magnetic field strength

> This quantity is taken as the *definition of the magnetic field strength B*
> when the current is *perpendicular to the field*
> $B = F/I\ell$
> (unit: N A^{-1} m^{-1} or tesla, T)
> or
>
> $F = BI\ell$

On a moving charge

A charge *q* moving in a B-field with velocity *v* at an angle θ to the field experiences a force given by:

$F = Bqv\sin\theta$

F

$+q$ → v

B

Remind yourself how this result is applied in an electric motor.

The direction of F is given by the L.H. rule – make sure you can apply it.

A magnetic field is commonly called a B-field.

If the current makes an angle θ with the field, F is multiplied by sin θ.

Be careful how you apply the L.H. rule if q is negative (conventional current is . . .).

Path of a moving charge in a magnetic field

> Since **F** is always perpendicular to **v**, a charge moving at 90 degrees to **B** must be bent into a *circular* path.
>
> The size of the circle is given by applying Newton's Second Law to the motion
> force = mass × acceleration
> $Bqv = m(v^2/r)$
> where r is the radius of the circle and v^2/r is the *centripetal* acceleration
>
> $r = mv/Bq$

Two consequences of this result:

- if it is rearranged to give $mv = Bqr$ the *momentum* can be found directly from the radius of the circle
- if it is rearranged to give $v/r = Bq/m$ the left-hand side is the *angular velocity* ω, showing that for a given charged particle in a fixed magnetic field the angular velocity is constant and so the time taken for a complete revolution is constant (principle of the *cyclotron*).

Example

A singly charged ion of mass 3.3×10^{-26} kg makes a circular track of radius 140 mm in a detector when a B-field of 1.3 T is applied perpendicular to the velocity. What is (a) the momentum and (b) the kinetic energy of the ion?
Charge $e = 1.6 \times 10^{-19}$ C .

Stage 1
Calculate momentum:

$$mv = Bqr$$
$$= 1.3\text{T} \times 1.6 \times 10^{-19}\text{C} \times 140 \times 10^{-3}\text{m}$$
$$= 2.9 \times 10^{-20} \text{ kg m s}^{-1}$$

Stage 2
Find the velocity first:

$$v = p/m \text{ where } p \text{ is the momentum}$$
$$= 2.9 \times 10^{-20} \text{ kg m s}^{-1}/3.3 \times 10^{-26} \text{ kg}$$
$$= 8.8 \times 10^5 \text{ m s}^{-1}$$

Stage 3
Now find kinetic energy:

$$E_k = mv^2/2 = 3.3 \times 10^{-26} \text{ kg} \times (8.8 \times 10^5 \text{ m s}^{-1})^2/2$$
$$= 1.3 \times 10^{-14} \text{ J (or 80 keV)}$$

Note also the useful relation which takes you directly from momentum to kinetic energy:

$$E_k = p^2/2m = (2.9 \times 10^{-20} \text{ kg m s}^{-1})^2/(2 \times 3.3 \times 10^{-26} \text{ kg})$$
$$= 1.3 \times 10^{-14} \text{ J}$$

Applications to particle physics and the mass spectrometer.

Magnetic flux, flux density and flux linkage

- If a closed loop of part of a circuit of area A (of any shape) is placed in a B-field so that the field is perpendicular to the plane of the area, the *flux* through A is:

$$\Phi = BA \text{ (unit: T m}^2 \text{ or weber, Wb)}$$

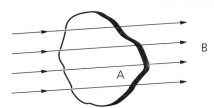

If B makes an angle θ with the plane of A Φ becomes $BA\sin\theta$ – flux is zero if $\theta = 0$.

- Rearranging, $B = \Phi/A$, giving an interpretation of B as *flux per unit area* or flux *density*.
- An alternative unit of B to the tesla is *weber/m²* (Wb m^{-2})
- If the loop has N turns of conductor:

the *flux linkage* through it is
$$N\Phi = BAN$$

Example
A solenoid has a cross-sectional area of 2×10^{-3} m^2 and in its central region there is a uniform B-field of 30 mT.

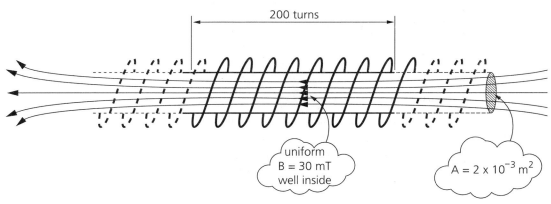

What is (a) the flux density in this region, (b) the flux and (c) the flux linkage through the central 200 turns?

(a) The flux density is the same as B but is usually expressed in Wb m^{-2} so is 30×10^{-3} Wb m^{-2} or 3.0×10^{-2} Wb m^{-2}

(b) $\Phi = BA = 3 \times 10^{-2}$ Wb m$^{-2} \times 2 \times 10^{-3}$ m$^2 = 6 \times 10^{-5}$ Wb

(c) $N\Phi = 200 \times 6 \times 10^{-5}$ Wb $= 0.012$ Wb
(sometimes this unit is written as weber-*turns* for flux *linkage*).

But the SI unit of flux linkage is still weber.

Electromagnetic induction

This is the production of an emf (and a current if there is a *complete* circuit) by the interaction between a conductor and a magnetic field.

There are *two distinct ways* of producing the emf (although in the end they are both included in one general overall statement).

Moving conductor, constant B-field ('dynamo' effect)

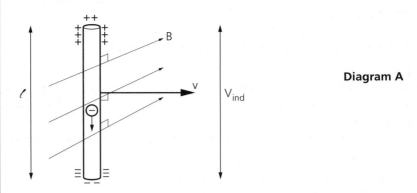

Diagram A

You need to understand the stages but not reproduce the theory in an exam.

- The wire moves perpendicular to its length *and* to the B-field, 'cutting' field lines.
- The free electrons in the wire move with the velocity of the wire.
- They are moving at 90 degrees to the B-field so they experience a force along the length of the wire.
- So they drift to one end, causing a charge separation.

This sets up an *E-field* in the wire resulting in a *potential difference* between the ends of the wire – the *induced emf* V_{ind}:

$$V_{ind} = B\ell v$$

The Hall effect arises from a rather similar argument – check on the principle of its application in the Hall probe for comparing and measuring B-fields.

Fixed conductor, varying B-field

- If the flux through a loop of circuit *changes* there will be an induced emf in the loop.
- The size of the emf depends on how *rapidly* the flux changes.
- If there is a change ΔB in field in a time interval Δt:

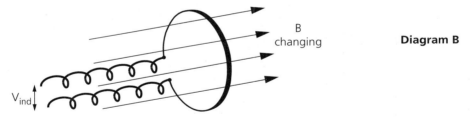

Diagram B

B changing

All transformers (where nothing actually moves) produce their output (secondary emf) in this way.

$V_{ind} = A(\Delta B/\Delta t)$ for *each* loop of circuit
or $V_{ind} = NA(\Delta B/\Delta t)$ for the whole circuit of N turns

$$V_{ind} = \Delta(N\Phi)/\Delta t$$
This is *Faraday's Law* of electromagnetic induction

Direction of induced emf is given by *Lenz's Law*:

'The direction of the induced emf tends to oppose the change causing it.'

If there is a complete circuit and an induced *current* flows then it will create an effect which *actually* opposes the change causing it.

Lenz's Law follows from the Law of Conservation of Energy. If it were not true then dynamos could keep turning, generating power without any external mechanical energy input.

This is the mean value over a time interval Δt – the calculus notation $d\Phi/dt$ gives the instantaneous value.

Examples of Lenz's Law

(They refer to the diagrams on page 78.)

Straight wire moving cutting field lines (diagram A)

- There is no induced current (no circuit).
- If the ends of the wire *were joined* to complete the circuit (by a stationary conductor) then there would be a conventional current flow *up* the wire (top of the wire behaves like a positive battery terminal).
- This induced current sets up a force (=$BI\ell$) on the wire to the left, opposing the motion.

Changing flux through a loop (diagram B)
- If the loop is part of a complete circuit a current will be induced in it.
- This current sets up *its own* magnetic field.
- If B is *increasing* then the magnetic field from the induced current is to the *left* (opposing change).
- If B is *decreasing* (but still in the *same* direction) the induced current sets up a magnetic field to the *right* (in the same direction as B).
- To decide the actual direction of induced current round the loop you will need to remind yourself of the link between current and field directions for a loop of current.

Example 1 (moving wire)
A wire of length 1.5 m moves perpendicular to itself with a velocity of 5.0 m s^{-1}. It is also at 90° to a B-field of 0.1 T. How much flux is cut in 5 seconds? What is the rate of cutting flux?

Stage 1
Put the information on a diagram and decide what *area* is involved (since flux always needs an area).

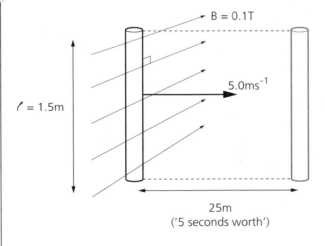

B = 0.1T

5.0ms^{-1}

ℓ = 1.5m

25m
('5 seconds worth')

> In 5 seconds the wire moves 25 m
> so the area required is
> the rectangle 1.5 m × 25 m

Stage 2
Calculation of flux:

> All the flux lines passing through this area have been 'cut'. Calling this $\Delta\Phi$
> $$\Delta\Phi = 0.1 \text{ T} \times (1.5 \text{ m} \times 25 \text{ m})$$
> $$= 3.75 \text{ T m}^2 \text{ (Wb)}$$

Textbooks may incorporate Lenz's Law into Faraday's by using a negative sign – interpreting this needs much care (and is often best ignored!).

Can you use a search coil for detecting and comparing alternating B-fields (at the same frequency)?

Extreme caution! Note that what is opposed is the change in flux, not the flux itself.

This is the area 'swept out' by the wire.

Remember
1 V = Wb s^{-1} so
this has to be
the induced
emf.

Faraday's Law
covers the
dynamo effect
as well.

This is a huge
field
– in air expect
only millitesla
fields.

Flux linkage per
unit current
$(N\Phi/I)$
is called the
self-inductance
L
here it is
$(22.5/10)$ Wb A^{-1}
or 2.25
henry (H).

If a large flux
linkage
(lots of iron!) is
destroyed very
quickly large and
potentially
dangerous
voltages can be
produced.

Check how
much detail
your syllabus
expects on
inductance.

In calculus
notation
LdI/dt for
instantaneous
values.

Stage 3
Appreciate that *rate* always means per second, so have to divide by a time interval:

> Here $\Delta t = 5$ s, so rate of cutting
> $\Delta\Phi/\Delta t = 3.75$ Wb/ 5 s
> $= 0.75$ V

Check:
Calculating *Blv* gives
$V_{ind} = 0.1$ T $\times$ 1.5 m $\times$ 5 m s^{-1} = 0.75 V
confirming that the 2 approaches do give identical results.

Example 2 (stationary circuit, changing flux)
A large electromagnet consists of 600 turns wrapped around an iron core of cross-sectional area 2.5×10^{-3} m^2. A current of 10 A produces a B-field in the iron of 15 T.
The current is suddenly switched off and drops to zero in 50 ms. Calculate the average induced voltage across the coil terminals and state any assumptions you have to make.

Stage 1
Calculate the flux linkage initially:

> $\Phi = BA = 15$ T $\times 2.5 \times 10^{-3}$ m^2 = 0.0375 Wb
> $N\Phi = 600 \times 0.0375$ Wb = 22.5 Wb (or Wb-turn)

Stage 2
Calculate the *change* in flux linkage:

> Assumption: when the current is zero the flux is zero – not true exactly since there
> will be some residual magnetisation in the iron.
> $\Delta(N\Phi)$ = final flux linkage – initial flux linkage
> $= 0 - 22.5$ Wb
> Taking the magnitude only (the minus sign shows a decrease)
> $\Delta(N\Phi) = 22.5$ Wb

Stage 3
Apply Faraday's Law:

> $V_{ind} = \Delta(N\Phi)/\Delta t = 22.5$ Wb/5.0×10^{-3} s = 4500 V

Self inductance

- An *inductor* is a component which creates flux *linking itself* when a current flows through it (the commonest example is a solenoid or coil).
- If the current changes the flux will change, inducing a voltage in the inductor itself (see previous example 2).
- This is called a *self-induced* voltage and by Lenz's Law must oppose the change in current (sometimes called a *back emf*).
- The magnitude of this self-induced voltage is given by:

> $V = L(\Delta I/\Delta t)$
> (mean value over a time interval Δt)
> where L is a property of the inductor called its
> *self-inductance*
> (SI unit: V s A^{-1}, Wb A^{-1}, henry (H))

Current growth in a circuit with an inductor and a resistor

- Such a circuit is said to be *inductive* and to possess the property of *inductance*.
- The inductor is normally taken to have no resistance of its own (a *pure* inductor).

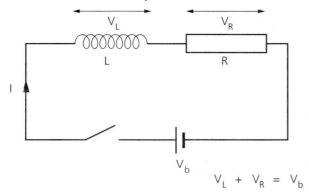

$$V_L + V_R = V_b$$

An inductive d.c. circuit

- When the battery is connected the current starts to grow.
- Which creates a self-induced voltage in the inductor.
- Which opposes the battery voltage and slows down the current growth.
- Initially the current is zero and its rate of growth is determined by L only ($= V_b/L$).
- Eventually the current reaches its maximum value given by the normal expression V/R – once the current has stopped changing the inductance has no effect.

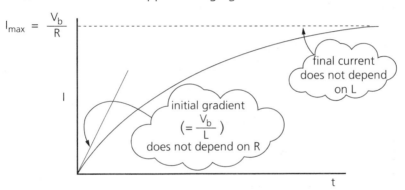

Current growth in an inductive circuit

B-fields produced by currents

Solenoid

Experiments carried out with a Hall probe to measure the B-field inside a solenoid with the following description:

- length ℓ which is long compared with the width of its cross-section (say at least 5 times – sometimes called 'infinitely long')
- wound with N turns and carrying a current I
- no medium inside it (strictly a vacuum, but air makes no perceptible difference),
produce the following results:
- shape and area of the cross-section have no effect on B
- provided the Hall probe is not too near an end (to within about 10% of its length) the value of B at the point of measurement is *independent of its position* (either along or perpendicular to the axis) – hence a *uniform field*
- $B \propto I$ (N and ℓ constant)
  $B \propto N$ (I and ℓ constant)
  $B \propto 1/\ell$ (N and I constant)

(The last relationship needs some care experimentally – you have to vary ℓ keeping N fixed, so use combinations of solenoids wound differently).
The proportionalities combine to give:
B = constant × (NI/ℓ)
(remember for a vacuum at this stage)

- The constant is a fundamental magnetic constant called the *permeability of free space* and is written as μ_0 (in practice it is just called 'mu-nought').
- It is in fact not an experimentally determined constant but follows from the definition of the ampere.

$$B = \mu_0(NI/\ell)$$
$$\text{where}$$
$$\mu_0 = 4\pi \times 10^{-7} \text{ T m A}^{-1} \text{ (exactly)}$$

Other shapes of circuit

Flat circular coil

The B-field at the centre of a flat circular coil, N turns radius r (like a very thin section of a solenoid) is given by:

$$B = \mu_0(NI/2r)$$

The field is not uniform and changes rapidly away from the centre
(along the axis and perpendicular to it)

Long straight wire

At a distance r from a straight wire ($r \ll$ length of wire) carrying current I:

$$B = \mu_0(I/2\pi r)$$

Force between wires

A pair of long parallel wires carrying currents I_1 and I_2 separated by distance r exert a force on each other.
- The wires attract if the currents are in the same direction.
- They repel if the currents are in the opposite direction.
- Combining the expressions for field strength from one of the wires and the force in a B-field ($F=BI\ell$) on the other wire gives the following result for the force *on a length ℓ* of either wire:

$F = \mu_0(I_1 I_2 \ell/2\pi r)$
- The SI definition of the ampere uses the following values:
$F = 2 \times 10^{-7}$ N, $\ell = r = 1$ m
when $I_1 = I_2 = 1$ A
The value of μ_0 ($4\pi \times 10^{-7}$ N A^{-2}) follows at once.

Significance of μ_0

μ_0 is important for 2 reasons:
- it provides the fundamental link between current and magnetic field strength – it always appears (multiplied by current) whenever a B-field has to be calculated
- Maxwell's electromagnetic theory calculates a theoretical value for c, the velocity of electromagnetic waves in a vacuum

2 parallel coils separated by their radius are called 'Helmholtz coils' – they produce a nearly uniform field in the space between them.

You do not need to memorise these formulae but check to see how much use you need to make of them.

Can you show 1 T m A^{-1} is the same as 1 N A^{-2}?

$$c = 1/\sqrt{(\varepsilon_0 \mu_0)}$$

so μ_0 is one of the components (ε_0 being the other one) of this fundamental constant of nature

If you can make the units of the RHS reduce to those of velocity you have a very good understanding of electric and magnetic fields!

Medium other than a vacuum

- The field inside a solenoid increases if a material substance forms the core.
- The increase is only slight for non-magnetic materials.
- But can be huge (several thousand times) for *ferromagnetic* (iron, cobalt or nickel based) materials.

Every material has its own *relative permeability* μ_r (a pure number) which multiplies μ_0 in the appropriate equation for B to give *the permeability* μ:

$$\mu = \mu_r \, \mu_0$$

Alternating current (AC)

- A dynamo coil rotating in a uniform magnetic field produces an alternating voltage given by a similar expression to that for SHM:

$V = V_0 \sin \omega t$
where V_0 is the maximum voltage (amplitude) and
$\omega = 2\pi \times$ frequency

Or cos ωt.

- If there is a complete circuit and current flows, the current varies similarly:

$I = I_0 \sin \omega t$

Assuming I and V are in phase.

- The angular velocity of the rotating coil is ω, so for 50 Hz mains AC the power station generator is rotating at 50 revolutions per second (3000 rpm).

Root mean square (RMS) values and power

- The mean value of I (or V) over 1 cycle must be zero (positive for half the time, negative for the other).
- This is no help in calculating power, since the power delivered by a source into a resistor cannot depend on the direction of the current.
- Taking current as an example, the procedure is to
  – square I first
  – find the mean value of I^2
  – take the square root of this value.
  The graphs show the process.

Look back to RMS molecular speeds – not of course sinusoidal!

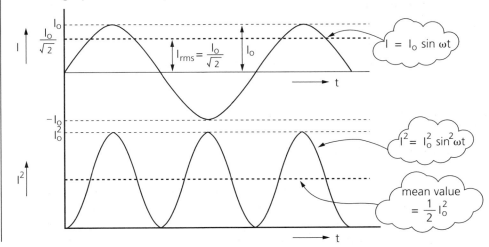

The mean value of I^2 is clearly $I_0^2/2$, so:

$$I_{RMS} = \frac{I_0}{\sqrt{2}}$$

and similarly for V

- So the peak value is the RMS value $\times \sqrt{2}$.
- The mean power dissipated as heat in a resistor by AC can be calculated exactly as for DC provided RMS values are used:

$$
\begin{aligned}
P &= V_{RMS} \times I_{RMS} \\
&= I_{RMS}^2 \times R \\
&= V_{RMS}^2/R
\end{aligned}
$$

- AC meters are generally calibrated to read RMS values directly.

Transformers

Principle of operation

An example of electromagnetic induction by changing flux – nothing actually moves.

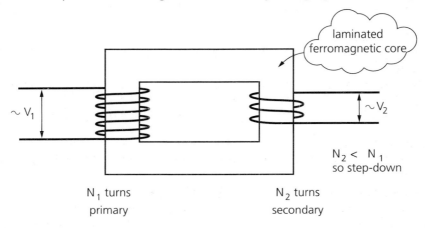

laminated
ferromagnetic core

$\sim V_1$

$\sim V_2$

$N_2 < N_1$
so step-down

N_1 turns
primary

N_2 turns
secondary

A transformer

- AC is passed through a coil wound on a ferromagnetic core (the *primary* current I_1) by applying the primary voltage V_1.
- I_1 creates an alternating flux in the core.
- Which links a second coil (the *secondary*) wound on the same core.
- The alternating flux induces an emf (obviously also alternating) in the secondary, V_2.
- At this stage, if the secondary is on open circuit (so that no secondary current flows) *no power is being delivered to the secondary* (and in principle therefore none drawn from the primary, if there were no losses).
- Finally, if the secondary is connected to a load, a current I_2 will flow in the secondary, the net effect being a transfer of power from primary to secondary via the magnetic link.

Output voltage and current

If the secondary is on open circuit, the following simple relation holds accurately:

$$V_2/V_1 = N_2/N_1$$

- So the voltage ratio is simply the turns ratio ($I_2 = 0$).
- A *step-up* transformer has $N_2 > N_1$, and so $V_2 > V_1$; a step-down is the other way.

- If the secondary is on load (drawing power with a current I_2) then if it were 100% efficient (laboratory ones won't be):
$V_1I_1 = V_2I_2$
so that:
$I_2/I_1 = V_1/V_2$
and finally:

> $I_2/I_1 = N_1/N_2$
> current is transformed in the opposite way to the voltage

- In practice this current-turns relationship holds quite accurately even at efficiencies well below 100%.

Uses of transformers

Safety
- There is no electrical connection between primary and secondary.
- So irrespective of any voltage change which might be operating, accidentally touching *one* of the output terminals (not both!) will not produce a shock (if the 'live' side only of the mains is touched there is the risk of electrocution).
- Such an arrangement is called an *isolating* transformer – used in hazardous situations like building sites and where water is involved (legal requirement in bathrooms).

Domestic

- Equipment designed to run off a battery or mains will be fitted with a step-down transformer.
- Which together with *rectification* and *smoothing* circuits make the output like that of a battery.
- Will produce an output of perhaps 6 volt (DC) from 240 volt (AC).

Power transmission
- Power distributed around the country will produce heat (power loss) in the transmission cables.
- This loss is proportional to the square of the current (I^2R).
- So it is more efficient to transmit a given power using as small a current as possible.
- This means at as *high a voltage* as possible (power delivered = transmission voltage x current).

- A step-up transformer is used at the output of a power station.
- And the voltage is correspondingly transformed down where the power is needed.

AC through capacitors and inductors

Capacitors

- The current and voltage have a phase difference of $\pi/2$ radians.
- The current reaches its maximum before the voltage (current *leads* voltage).
- The mean power delivered to a capacitor is zero.
- The peak (and RMS) values are related by:

> $V_0/I_0 = 1/\omega C$
> This ratio is called the *capacitive reactance* X_c
> and is measured in ohms

- The reactance decreases as the frequency increases, so capacitors conduct *better* at higher frequencies.

Inductors

- The current and voltage also have a phase difference of $\pi/2$ radians.

- But the current reaches its maximum after the voltage (current *lags* voltage).
- The peak (and RMS) values are also linked by a reactance:

$$V_0/I_0 = \omega L$$
which is called the *inductive* reactance X_L
and is measured in ohms

Are you expected to be able to draw graphs of I-t and V-t for C and L in the exam?

- The reactance increases as the frequency increases so inductors conduct *less well* at higher frequencies.

Inductor-capacitor series combination

- Because of the different phase relations between current and voltage for the two components, when they are in series their reactances *subtract* to give the overall reactance of the circuit:

$$X_{tot} = X_L - X_C$$

- The sign of X_{tot} does not matter, just its magnitude.

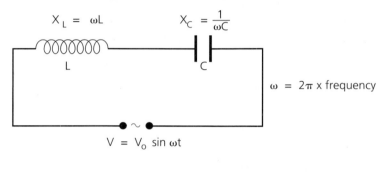

$$X_{tot} = \omega L - \frac{1}{\omega C}$$

Resistance-reactance series combination

Reactance and resistance 'add' like vectors at 90° - they are sometimes called 'phasors'.

- Because of the phase relations (and despite their both being measured in ohms), resistance and reactance do not simply add together.
- The combined effect of R and X_{tot} is to produce an effective 'overall resistance to AC' called the *impedance Z* (measured in ohms):

$$Z^2 = R^2 + X_{tot}^2$$

Minimum series impedance and resonance

- Z is a minimum when the total reactance is zero (and Z is then just $= R$).
- This occurs when the frequency is such that:

$$\omega L = 1/\omega C$$
or
$$f = 1/\{2\pi\sqrt{(LC)}\}$$

- For a given voltage amplitude this frequency is the one to give the biggest current amplitude and the circuit will *resonate* (see Chap. 4) at this frequency.
- The circuit is capable of free (but normally damped) natural oscillations at this frequency.

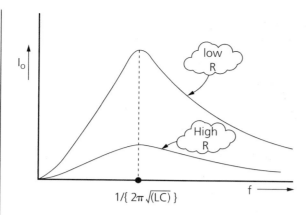

Series resonance response

Example
(a) What are the reactances at 50 Hz of a 0.10 μF capacitor and a 10 H inductor?
(b) How do the reactances change for a frequency of 160 Hz?
(c) What is the effect of combining them in series at this frequency?

Stage 1
Calculate the ω value:

> (data to 2 sig. figs. so results to the same accuracy)
> $\omega = 2\pi \times 50$ Hz
> $= 310$ s^{-1}

Stage 2
Use the reactance relations:

> $X_c = 1/\omega C = 1/(310$ s$^{-1} \times 0.10 \times 10^{-6}$ F)
> $= 3.2 \times 10^4\ \Omega$
>
> $X_L = \omega L = 310$ s$^{-1} \times 10$ H
> $= 3.1 \times 10^3\ \Omega$

Stage 3
Either re-calculate from scratch or note that the frequency has increased by a factor $160/50 = 3.2$.
In this stage it will be used as a scaling factor (up or down as appropriate) for each reactance:

> $X_c = 3.2 \times 10^4\ \Omega/3.2$ (lower reactance)
> $= 1.0 \times 10^4\ \Omega$
>
> $X_L = 3.1 \times 10^3\ \Omega \times 3.2$ (higher reactance)
> $= 0.99 \times 10^4\ \Omega$

Stage 4

> The two reactances are almost identical, producing almost zero total reactance (remember they have to be subtracted), so the circuit is almost at resonance.

8 Nuclear physics

Rutherford model of the atom

- All the positive charge in an atom (carried by the *protons*) is located in the *nucleus*.
- Practically all the mass of an atom is in the nucleus (proton + *neutron* mass).
- The electrons (equal in number to the protons) orbit the nucleus in a sequence of shells.
- The effective volume of an atom is determined by the radius of the outer shell ($\approx 10^{-10}$ m).

Evidence for this model and for the size of the nucleus is provided by the
Geiger Marsden scattering experiment.
- Alpha (α) *particles* (see below) were fired at very thin gold foil.

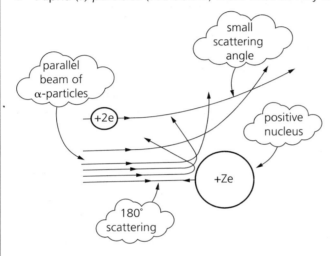

- The majority went straight through with very little deviation.
- A small proportion (but much bigger than predicted by a uniform model of an atom) were deflected back through very large angles.
- These large deflections were interpreted as near direct hits and subsequent repulsion between the positive α and a very small concentration of positive charge at the atom's centre.
- Careful measurements of the proportion scattered at different angles showed that the force of repulsion followed Coulomb's Law ($F = (1/4\pi\varepsilon_0)Q_1Q_2/r^2$) right down to separation distances of the order of 10^{-14} m which gives an indication of nuclear radius.

As a **consequence** of this model:

- practically all of matter in its normal state consists of empty space
- the density of *nucleons* (neutrons or protons) is inconceivably large ($\approx 10^{16}$ kg m^{-3}).

Example
An α-particle of kinetic energy 5.0 MeV makes a head-on collision with a gold nucleus and returns along its original path. Gold has 79 protons in its nucleus.

(a) describe the energy changes involved
(b) hence find the closest (centre-to-centre) distance of the two particles

Stage 1
Put the information on a labelled diagram:

High energy scattering shows departures from Coulomb's Law as the field of the short-range strong nuclear force is entered.

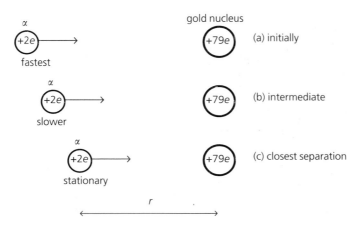

Stage 2
Consider kinetic and potential energies:

> In (a)$E_k = 5.0$ MeV $(= 5 \times 10^6 \times 1.6 \times 10^{-19}$ J)
> $E_p = 0$ (infinite separation)
>
> In (b)E_p has increased (separation decreased) and so E_k has reduced
>
> In (c)$E_k = 0$ (stationary for an instant)
>
> $E_p = (1/4\pi\varepsilon_0)\ Q_\alpha Q_{Au}/r$ (the maximum value)

Stage 3
Apply the Law of Conservation of Energy:

> The total energy at every stage is the same
>
> Comparing (a) and (c)
>
> $5 \times 10^6 \times 1.6 \times 10^{-19}$ J $= 9 \times 10^9$ N m^2 C$^{-2} \times (+79)(+2) \times (1.6 \times 10^{-19}$ C$)^2$/r
>
> r $= 3.64 \times 10^{-26}$ N m^2/8.0 $\times 10^{-13}$ J $= 4.6 \times 10^{-14}$ m

This is applying the law in the form $E_k + E_p$=constant with the zero energy terms omitted.

Remember that 1 J = 1 N m.

But the nucleus by itself has no chemical properties.

Nuclear formulae

- The chemical symbol of the element is used.
- The nucleon number (or mass number) A is the total number of nucleons in the nucleus.
- The proton number (or charge or atomic number) Z is the number of protons in the nucleus.

> a particular *nuclide* (specific combination of A and Z)
> is written in the notation:
>
> nucleon number
> proton number
> chemical symbol
>
> $_Z^A X$
>
> 2 nuclides with the same Z but different A (and therefore the same element) are *isotopes* of the element

$^{7}_{3}$Li and $^{9}_{4}$Be are nuclides.

- The nuclides $^{7}_{3}$Li and $^{6}_{3}$Li are isotopes of lithium (with 4 and 3 neutrons respectively).
- The neutron, proton and a γ-ray are written $^{1}_{0}$n, $^{1}_{1}$H (or $^{1}_{1}$p) and $^{0}_{0}\gamma$

The α-particle is the nucleus of the common isotope of helium so is often written $^{4}_{2}$He.

> **Nuclear masses** are measured in the *unified atomic mass unit (u)* which is one-twelfth of the mass of the *atom* (not nucleus) of the carbon isotope $^{12}_{6}$C.
>
> $1\ u = 1.66 \times 10^{-27}$ kg
>
> mass of proton $(m_p) \approx$ mass of neutron $(m_n) \approx 1$ u

Radioactivity

- Is the general name for a variety of spontaneous processes (usually called *decay*) which may happen in the nucleus where the resulting nucleus is a different nuclide from the original one.
- The *evidence* that this is happening is the emission of various types of radiation which can be detected.

A common exam question is to balance values of A and Z in a decay or chain of decays.

The decay processes

---

alpha (α) decay

- increases the stability of more massive nuclei (Z above about 80)
- the nucleus ejects at high velocity a tightly bound group of 4 particles – 2n and 2p (the *alpha particle*) – written

  $^{4}_{2}\alpha$ or $^{4}_{2}$He

- the resulting nucleus (the *daughter*) has

  – A reduced by 4
  – Z reduced by 2

- the daughter nucleus is often in a high energy state (*excited*) and can remove more energy by emitting a pulse of *gamma (γ)* radiation (an electromagnetic wave of very short wavelength)

  example:
  $^{226}_{88}$Ra $\rightarrow$ $^{222}_{86}$Rn $+ ^{4}_{2}$He

---

The radon daughter nuclide will emit γ-radiation and is also itself α-emitting.

---

beta (β) decay

- often occurs in lighter nuclei when the number of neutrons is greater than the number of protons
- increases the stability of these nuclei by ejecting a fast *electron* (the *beta* particle) from the nucleus, usually written:

  $^{0}_{-1}\beta$ or $^{0}_{-1}$e

- the electron results from the decay of a neutron into a proton and electron
- the daughter nucleus has

  – A unchanged
  – Z increased by 1

  example:
  $^{90}_{38}$Sr $\rightarrow$ $^{90}_{39}$Y $+ ^{0}_{-1}\beta$

Strictly this is β^{-} decay – there is also the less common β^{+} (positron) emission where a proton decays to a neutron.

Detection and identification of the radiation

- Detection is usually by the ability to cause ionisation in a gas (cloud chamber or Geiger-Müller tube).
- Identification of each type is by:
  – different penetrating powers through absorbers
  – different deflection in a magnetic field.

Since α and β are charged moving particles a B-field can bend them into circular paths (in different directions).

A summary of the radiation properties

	alpha	beta	gamma
approx. energy/MeV	5	0–0.5	up to 1 (per pulse)
ionising power	very strong	medium	weak
penetrating power (stopped by)	thin paper or 5–10 cm air	few mm Al	many cm lead or m of concrete
deflection by B-field	very small	large (opp. to α)	zero

Background radiation

There is a continual background of radiation which is important for 2 reasons:

- in laboratory work it has to be allowed for by measuring it in the absence of a source and then subtracting its activity from measured activities with a source
- our bodies are continually exposed to it and there is a potential health hazard.

Sources of background (all types of radiation including X-rays)

Typical average exposure is:

- cosmic radiation from space (about 10%)
- medical diagnosis and treatment (about 12%)
- naturally occurring radioactivity in building materials, the ground and food (about 25%) but *not including*
- Radon (a gas, a decay product of uranium in rocks) which is present in the soil and air: it and its decay products are both alpha and beta active (about 47%).

Hazards and safety precautions

The radiation causes ionisation as it penetrates matter and all types may cause

- body cell damage
- DNA damage, possible leading to cancer or genetic disorders.

From sources *outside* the body, *gamma* is the most hazardous since it is likely to penetrate deeply into the interior of the body.

The greatest risk from *alpha* is from sources *inside* the body (perhaps from inhaling radon and its products); external alpha radiation is stopped by the surface layer of dead skin.

The 3 key components of a set of safety precautions and guidelines are:

- time – reduce overall exposure time to a minimum – the damage is cumulative
- distance – intensity decreases with distance from the source, and for gamma radiation it follows an inverse square law (as for other point sources)
- shielding – knowledge of the relation between energy and penetration for specific media for the different types of radiation is vital in designing shielding.

Uses of radioactivity

Here is a short reminder list of main applications of *radioisotopes* – if your syllabus or option requires more detailed knowledge you must refer to your source book.

- **Tracers** for tracking:

- movement and quantity of body and other fluids
- underground pipe leaks
- root uptake in plants
- **Crack testing** – gamma radiation is used in *non-destructive* testing for cracks in metals without dismantling machinery: particularly used in the aerospace industry.
- **Radioactive dating** – the half-life of ^{14}C is about 5700 years so this isotope can be used by archaeologists for dating back to about 30 000 years; longer lived isotopes can be used for dating over much longer periods.
- **Cancer treatment** – high energy gamma radiation from an external source of ^{60}Co will destroy cancer cells (*radiotherapy*).
- **Body imaging** – tracer isotopes, gamma radiation cameras and more recently Positron Emission Tomography (PET) can be used in conjunction with computer techniques to image sections of body organs.

Random nature of decay

- Radioactivity is a *random* process – it is governed by probability.
- All nuclei of a *specific nuclide* have the same probablitiy of decay *per unit time*.
- The probability of decay per second is called the *decay constant* λ (unit s^{-1})

Law of decay

The probability of decay is represented by the *fraction of a population* decaying in the time interval considered.

If from a population N a number ΔN decay in a time Δt then:

This can be rearranged to:

$$\Delta N/\Delta t = -\lambda N$$

(provided Δt is small compared with $1/\lambda$)

In calculus notation $dN/dt = -\lambda N$.

- $\Delta N/\Delta t$, the rate of decay, is called the *activity* (1 decay per second is an activity of 1 becquerel (Bq): 1 Bq = 1 s^{-1}).
- The activity is proportional to the population, so a graph of activity (which is what can actually be measured) against time will have exactly the same properties as one of population against time.
- It is usually impossible to measure N directly, but a knowledge of activity and λ enables N to be calculated.

Since the rate of decay is proportional to the population the decay must be *exponential*:

If N_0 is the population at $t = 0$ and N is the population at a time t

$$N = N_0 e^{-\lambda t}$$

The decay is often specified by the half-life T – the time taken for the population (or activity) to halve, where:

$$\lambda T = \ln 2$$
$$\text{so } T = \ln 2/\lambda = 0.693/\lambda$$

'ln' stands for $\log_e$' – it is also how your calculator is labelled, make sure you can use it!
See stage 2 example 2 following.

You need to understand how the graph of $\ln N$ (or ln of activity) against t is linear with gradient $(-\lambda)$ – often the best route into finding T experimentally.

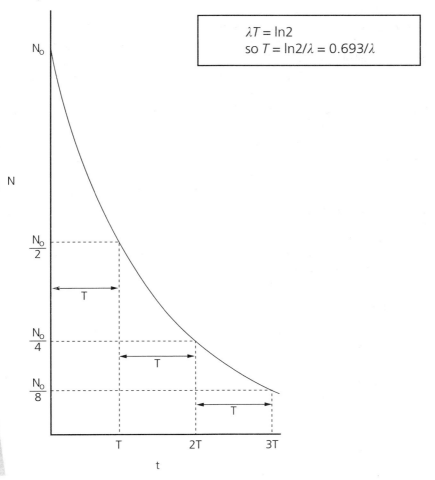

It is possible to use $N = N_0 e^{-\lambda t}$ but unless you are really expert at the algebra of exponentials and logs it is best avoided in an exam.

Example 1

A radioactive source has an initial activity of 1000 Bq and a half-life of 10.0 minutes. What is the activity after (a) 30 minutes and (b) 35 minutes?

Stage 1 for (a)

Before launching into a complicated calculation look at the time interval (30 min) in relation to the half-life (10 min) and register that you are dealing with 3 half-lives.

> After 1 half-life the activity has reduced by a fraction 1/2
> After 2 half-lives the activity has reduced by $(1/2)^2$ i.e. 1/4
> After 3 half-lives the activity has reduced by $(1/2)^3$ i.e 1/8
>
> So after 30 minutes the activity is 1000 Bq/8 = 125 Bq

Stage 2 for (b)

> Extending the idea of Stage 1, after n half-lives the activity has reduced to a fraction $(1/2)^n$ of its initial value.
>
> Here $n = 35/10 = 3.5$, so the fraction is $(1/2)^{3.5} = 0.088$ (or 1/11.3) and the activity is 1000 Bq /11.3 = 88 Bq

Instead use the method shown here, even when the number of half-lives is not a whole number.

Example 2

A sample of β-active $^{90}_{38}$Sr has an activity of 1.0×10^4 Bq. What is the mass of the isotope? It has a half-life of 28 years. The Avogadro constant is 6.2×10^{23} mol^{-1} and 1 year is 3.2×10^7 s.

Stage 1

> Identify from the data which of the decay relations to use.
>
> The data on the Avogadro constant and the requirement for *mass* suggest that numbers of nuclei are involved. The data on activity confirms that $\Delta N/\Delta t = -\lambda N$ will be used.
>
> λ is not given but the half-life is, so $\lambda T = \ln 2$ is also needed.

Stage 2
Calculate λ:

$$\lambda T = \ln 2$$
$$\lambda \times (28 \times 3.2 \times 10^7 \text{ s}) = 0.693$$
$$\lambda = 7.7 \times 10^{-10} \text{ s}^{-1}$$

Since activity is in Bq(s^{-1}) λ needs to be in s^{-1} so convert T from y→s.

Stage 3
Calculate N:

$$\Delta N/\Delta t = -\lambda N$$
$$-1.0 \times 10^4 \text{ s}^{-1} = -7.7 \times 10^{-10} \text{ s}^{-1} \times N$$
$$N = 1.3 \times 10^{13}$$

$\Delta N/\Delta t$ is negative since the population is decaying.

This is the number of nuclei of this isotope of Sr and is also the number of atoms.

Stage 4
Convert number of nuclei to mass:

> From the value of A (=90) the molar mass is 90 grammes = 0.090 kg
>
> 6.0×10^{23} atoms have a mass of 0.090 kg
> 1.3×10^{13} atoms have a mass of 0.090 kg $\times (1.3 \times 10^{13}/6.0 \times 10^{23})$
> $= 1.9 \times 10^{-12}$ kg

Be prepared to use 'hidden' data in the nuclear formula (as A here).

Energy in the nucleus

Binding energy

Suppose the α-particle could be pulled apart into its 4 separate nucleons against the strong nuclear attractive force:

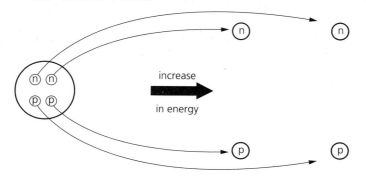

increase in energy

- Work has to be done to separate them so there is an *increase* in potential energy.
- So the single α-particle has *less* energy than the 4 constituent particles.
- The 4 separate particles have zero total potential energy (infinite separation).
- So the potential energy of the α-particle (or He nucleus) is *negative*.
- The energy is called the *nuclear binding energy* (B.E.): all nuclei possess this energy and it is *always* negative.
- It can be considered also as the (positive) energy needed to separate the nucleons to an infinite separation.

Mass change

Note that it is necessary to work with more sig. figs. than usual here – otherwise the differences vanish!

The mass of $^4_2$He is 4.0015 u.
The mass of 2 neutrons is 2.0173 u and for 2 protons is 2.0146 u giving a total mass for the 4 separate nucleons of 4.0319 u.

- The 4 separate nucleons have a mass *greater* than when bound in the nucleus by an amount 0.0304 u.
- They also have more energy.

This is an example of Einstein's *mass-energy equivalence* relation which is usually written as
$E = mc^2$
where c is the velocity of light in a vacuum ($= 3.0 \times 10^8$ ms^{-1}).

$E=mc^2$ applies to all energy processes not just nuclear.

It is more usefully written here as:

$$\Delta E = c^2 \Delta m$$

Whenever there is an energy change to a system ΔE there is a corresponding mass change Δm.

To work numerically with this relation, SI units have to be used.

Example
What is the binding energy of the $^4_2$He nucleus? 1 u $= 1.66 \times 10^{-27}$ kg

For this nucleus $\Delta m = 0.0304$ u
$= 0.0304 \times 1.66 \times 10^{-27}$ kg $= 5.05 \times 10^{-29}$ kg

$\Delta E = c^2 \Delta m = (3.0 \times 10^8$ ms$^{-1})^2 \times 5.05 \times 10^{-29}$ kg $= 4.54 \times 10^{-12}$ J

So the binding energy of this nucleus is -4.54×10^{-12} J or -28.4 MeV
(1 eV $= 1.6 \times 10^{-19}$ J)

Binding energy per nucleon

- For $^4_2$He the B.E. is -28.4 MeV shared by 4 nucleons, so the B.E. *per nucleon* is $-28.4/4$ or -7.1 MeV.
- This value is *approximately the same* for nearly all nuclei.
- But within the range of known nuclides there are some very important differences from this nearly constant value.
- These differences lead to the prediction of nuclear *fission* and nuclear *fusion* as energy liberating processes.

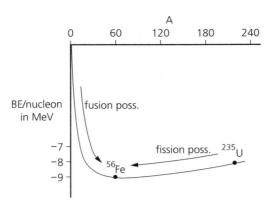

Predictions from the graph of B.E. per nucleon against mass number

- For massive nuclei to the right of the minimum at ^{56}Fe *fission* may occur: nucleons in a large nucleus may re-group themselves into 2 smaller nuclei with lower total energy.
- For light nuclei to the left of ^{56}Fe *fusion* may occur: nucleons in 2 light nuclei may join together to make a larger nucleus with lower total energy.
- Fission can be spontaneous although it is usually triggered by the addition of a neutron, as in a *chain reaction*.
- Fusion has to be started at very high temperatures ($\sim 10^8$ K) in order for the fusing nuclei to have sufficient KE (working against the electrostatic repulsion force) to come close enough together.

Example

In the fission reaction

$$^{235}_{92}U + ^{1}_{0}n \rightarrow ^{96}_{37}Rb + ^{138}_{55}Cs + 2^{1}_{0}n$$

how much energy is available (a) per fission (in eV) (b) per kg of ^{235}U?

Stage 1

Use a reference table to find the mass change:

> before fission:^{235}**U**:235.044 u   **n**:1.009 u
> **total mass:**236.053 u
>
> after fission:^{96}**Rb**:95.933 u ^{138}**Cs**:137.920 u   **2n:**2.017 u
> **total mass:**235.870 u
>
> $\Delta m = (235.870 - 236.053)$ u $= -0.183$ u (final minus initial)
> $\Delta m = -3.04 \times 10^{-28}$ kg

Stage 2

Convert Δm to energy change ΔE:

> $\Delta E = c^2 \Delta m = (3.0 \times 10^8 \text{ ms}^{-1})^2 \times (-3.04 \times 10^{-28} \text{ kg}) = -2.7 \times 10^{-11}$ J
> Since 1 eV $= 1.6 \times 10^{-19}$ J, $\Delta E = -170$ MeV

Stage 3

> Mass of 1 nucleus of uranium $= 235.044$ u $= 3.93 \times 10^{-25}$ kg
> Number of nuclei per kilogramme must be the inverse of this, i.e.
> $(3.93 \times 10^{-25} \text{ kg})^{-1} = 2.55 \times 10^{24}$ kg^{-1}
> So total energy available per kilogramme must equal
> energy per fission × fissions per kilogramme
> $E = 2.7 \times 10^{-11}$ J $\times 2.55 \times 10^{24}$ kg$^{-1} = 6.9 \times 10^{13}$ J kg^{-1}

Systems try to lower their overall PE.

Slight rounding error in neutron mass explains why 2.017 not 2.018 – be prepared for this sort of thing which is inevitable in deciding what accuracy to quote.

Δm is negative so there is a drop in energy – taken up as KE of fission fragments and eventually by collisions with surrounding atoms, raising the temperature.

There should be final rounding to a sensible no. of sig. figs.

9 Quantum effects

The photon

Diffraction and interference indicate the wave nature of electromag. radiation.

- Is a packet of electromagnetic energy sometimes considered as a particle.
- Travels at the velocity of light in the medium.
- Has energy proportional to the frequency of the electromagnetic wave.
- The number per second arriving at a point is a measure of the intensity.
- The intensity at any point is the photon energy multiplied by the number arriving per unit area per unit time.

Remember intensity is power per unit area.

> photon energy $E_{ph} = hf$
> where h is **Planck's Constant** ($= 6.63 \times 10^{-34}$ J s)
> and f is the frequency

Evidence for the existence and energy of photons comes from the way electromagnetic radiation interacts with matter:

- absorption – the *photoelectric effect* with metals
- emission – existence of *line spectra* from atoms.

Example
How many photons are emitted per second from a mercury vapour lamp working at 0.1 W if the wavelength is 250 nm (in the UV)?

Stage 1
Convert the wavelength to frequency:

Hz is the same as s^{-1}.

> $c = f\lambda$
> $f = c/\lambda$
> $= 3.0 \times 10^8$ m s^{-1}/250 $\times 10^{-9}$ m
> $= 1.2 \times 10^{15}$ Hz

Stage 2
Find the energy of 1 photon:

> $E_{ph} = hf$
> $= 6.6 \times 10^{-34}$ J s $\times 1.2 \times 10^{15}$ Hz
> $= 7.9 \times 10^{-19}$ J

Stage 3
Relate photon energy to energy per second radiated by lamp:

Rate of emission must have s^{-1} as its unit.

> No. of photons per second $= \dfrac{\text{energy per second}}{\text{energy per photon}}$
> $= 0.1$ J s^{-1}/7.9 $\times 10^{-19}$ J
> $= 1.3 \times 10^{17}$ s^{-1}

Photoelectric effect

- Is the emission of electrons from a metal surface when light shines on it.
- Is modelled as one photon colliding with one surface electron and handing over all its energy.

- The energy received by an electron:
  – gives it PE to cross the surface (the *work function W* for a metal)
  – and the balance is KE carried away from the surface.

The energy balance is given by
Einstein's photoelectric equation

$$E_{ph} = W + E_k$$

$$hf = W + E_k$$

If $hf < W$ then no emission is possible since the photon does not have enough energy to liberate the electron. The frequency for this condition is called the *cut-off* or (*threshold*) *frequency* f_c.

$f_c = W/h$

Changing frequency and brightness

- If $f \le f_c$ no emission is possible *however bright the light*.
- Increasing the brightness when $f > f_c$ increases the rate of emission of electrons (the *photoelectric current*) since more photons are available per second.
- The brightness has *no effect* on the KE.
- For a given surface the KE depends *only on frequency* and *increases linearly* with it.

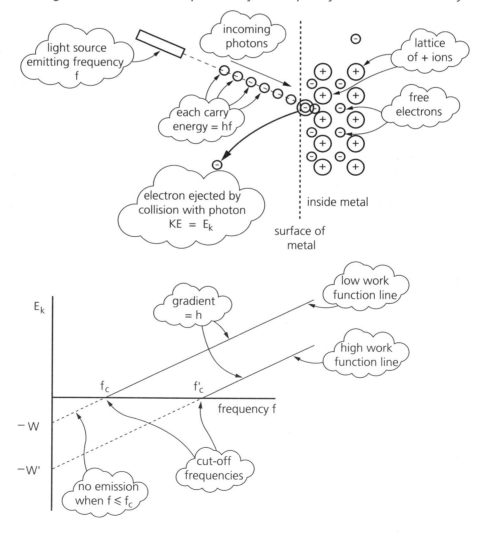

Kinetic energy

Kinetic energy of the ejected electrons is measured by:

- surrounding the photo-emitting metal by a grid
- applying a *negative* potential V_s to the grid relative to the metal (the *stopping potential*)
- increasing V_s until a critical value is reached which just stops the p.e. current.

When the stopping potential is reached:

$E_k = eV_s$

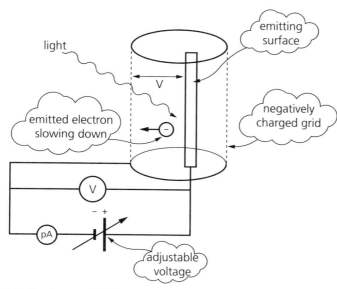

Emission of light

- A hot object like a lamp filament emits a *continuous spectrum* (all wavelengths).
- Light from a *gaseous element* (sodium *vapour*, mercury *vapour*, neon *gas* etc.) emits *specific wavelengths only*, called a *line spectrum*. Each wavelength is called a spectral line.

Atomic energy levels and emission of photons

- Each line has a very well defined frequency.
- So the light consists of photons of *certain energies only* – the light energy is *quantised*.
- The energy within the atom is also quantised – only specific energies are allowed on a ladder of levels associated with electron orbits.
- A photon is emitted when an atom in a high energy state (an *excited* state) drops to a lower level.
- The photon energy is equal to the difference in energy levels ΔE:

$$\Delta E = hf$$

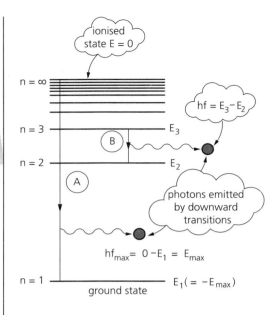

n = ∞   ionised state E = 0

n = 3 — E₃

B

hf = E₃–E₂

photons emitted by downward transitions

n = 2 — E₂

A

$hf_{max} = 0 - E_1 = E_{max}$

n = 1 —————— $E_1 (= -E_{max})$
ground state

energy levels are labelled by the value of
n - the principal quantum number

n = 1 is the ground state (lowest)

n = 2 upwards are the excited states

A typical atomic energy level diagram showing transitions and the emission of photons

- The highest energy state (ionisation) is given the value zero.
- All other energies must be *negative*.
- Transition A is the largest energy change possible and emits the highest frequency photon in the spectrum.
- Transition B is a typical jump from *third quantum level* (n = 3) to the second, emitting a photon of energy $E_3 - E_2$.

> Over any one very short time interval an atom will only be emitting *one photon*
>
> The observed line spectrum with many frequencies results from the combined effect of *all the atoms in the gas*

Example

A wavelength in the UV part of the hydrogen spectrum is 122 nm and is known to be due to the n = 2 to n = 1 transition. If the ground state energy is −13.6 eV what is the energy of the n = 2 level?

Stage 1
Calculate the photon energy:

> wavelength must be converted to frequency
> $$f = c/\lambda$$
> $$= 3.00 \times 10^8 \text{ m s}^{-1}/122 \times 10^{-9} \text{ m}$$
> $$= 2.46 \times 10^{15} \text{ Hz}$$
> $$E_{ph} = hf$$
> $$= 6.63 \times 10^{-34} \text{ Js} \times 2.46 \times 10^{15} \text{ Hz}$$
> $$= 1.63 \times 10^{-18} \text{ J}$$

Stage 2
Convert to eV and relate to energy levels:

$$1.63 \times 10^{-18} \text{ J} = 1.63 \times 10^{-18} \text{ J}/1.60 \times 10^{-19} \text{ J eV}^{-1}$$
$$= 10.2 \text{ eV}$$

$$E_{ph} = E_2 - E_1$$
$$10.2 \text{ eV} = E_2 - (-13.6 \text{ eV})$$
$$E_2 = -3.4 \text{ eV}$$

Particles as waves

de Broglie's hypothesis

a particle with momentum p has an associated wavelength λ given by
$$\lambda = h/p$$
this wavelength is called the *de Broglie wavelength*

- Because h is very small, only atomic-sized particles will have sufficiently small momentum for λ to be detectable.
- If the de Broglie wavelength for an electron is comparable to atomic dimensions then *electron diffraction* effects can be observed (principle of electron microscope).
- If the de Broglie wavelength for an electron is comparable to *nuclear* dimensions (very high energy) then diffraction by the nucleus can be observed giving information on nuclear size.

Example 1
What is the de Broglie wavelength for an electron accelerated through a p.d. of 5.0 kV?
$m_e = 9.1 \times 10^{-31}$ kg

Stage 1
Apply the Law of Conservation of Energy to the accelerated motion:

Method 1
$$1/2 \, mv^2 = eV$$
$$v = \sqrt{(2eV/m)}$$
$$= \sqrt{(2 \times 1.6 \times 10^{-19} \text{ C} \times 5000 \text{ V}/9.1 \times 10^{-31} \text{ kg})}$$
$$= 4.2 \times 10^7 \text{ m s}^{-1}$$
$$p = mv = 3.8 \times 10^{-23} \text{ kg m s}^{-1} \text{ (or N s)}$$

Method 2
$$E_k = p^2/2m \text{ and } E_k = eV$$
$$p = \sqrt{(2meV)}$$
$$= 3.8 \times 10^{-23} \text{ kg m s}^{-1} \text{ (or N s)}$$

Stage 2
Apply the de Broglie relation:

$$\lambda = h/p$$
$$= 6.6 \times 10^{-34} \text{ J s}/3.8 \times 10^{-23} \text{ N s}$$
$$= 1.7 \times 10^{-11} \text{ m}$$

Example 2

The electron beam of example 1 is fired in a vacuum through a thin slice of graphite. The regularly spaced rows of carbon atoms make the crystal lattice behave like a diffraction grating to the electron beam, the diffraction following the ordinary rule for waves ($n\lambda = s \sin \theta_n$) using the de Broglie wavelength. If the atomic spacing is 1.2×10^{-10} m at what angle does the 1st order beam emerge?

> The de Broglie wavelength has already been found to be
> 1.7×10^{-11} m
> so using the grating relation
> $1 \times (1.7 \times 10^{-11} \text{ m}) = 1.2 \times 10^{-10} \text{ m} \times \sin \theta_1$
> $\theta_1 = 8.1°$

If it hits a screen 140 mm from the graphite describe what is seen on the screen.

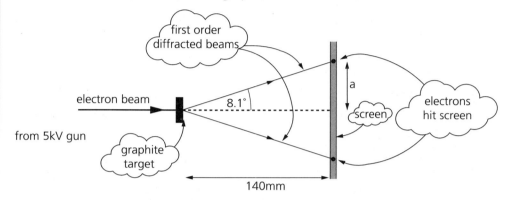

Electron diffraction by graphite

$a = 140$ mm $\times \tan 8.1°$ $\qquad = 20$ mm (2 sig.fig.)

Because the graphite sample will have many layers of atoms at all possible orientations (polycrystalline) there will be many diffracted beams at the angle 8.1°, forming an emerging cone of electrons. They hit the screen making a circle of radius 20 mm.

Wave-particle duality

This dual nature of matter and radiation – some phenomena best described on a wave basis using ideas of interference, diffraction and wavelength and others on a particle basis using momentum and kinetic energy – is called *wave-particle duality*. It forms the basis of Quantum Mechanics, the form of physical laws best suited to describe atomic and nuclear phenomena.

The wave nature of a moving particle can be applied to the electron in a hydrogen atom, where the electron is represented as a standing wave whose wavelength in the fundamental mode is twice the atom's diameter. This model, where the electron wave is thought of as representing the probability of finding an electron at particular distances, is remarkably successful at predicting the energy levels in hydrogen.

Questions

Where necessary use the following value:
$g = 9.8 \text{ N kg}^{-1}$ or 9.8 m s^{-2}

1 A ball is dropped from rest on to the floor and it bounces back to one quarter of its original height. Sketch a velocity–time graph for the motion, labelling its main features.

   (a) How does the graph show that the rebound height is one quarter of the drop height?

   (b) How does the graph show that the rebound *time* and rebound *speed* are both one half of the original values?

2 A bogged down vehicle is being recovered by two forces: one is 6000 N due north and the other 4500 N due east. What is the magnitude and direction of the resultant force vector?

3 A 5.0 kg mass is hung from a cord in which a spring balance has been inserted to measure the tension in the cord. What will the balance read?
The mass is now pulled sideways by a horizontal force F so that the cord makes an angle of 30° with the vertical.

   (a) What will the spring balance read? (Resolve vertically)

   (b) How large is F? (Resolve horizontally)

   (c) Repeat for angles of 20° and 10°. What do you notice about the change in F?

4 A missile is projected with a speed of 30 ms^{-1} from ground level at an angle of 60° to the horizontal. Find the values of:

   (a) The vertical component of velocity at launch.

   (b) The time to reach its maximum height (vertical velocity component zero).

   (c) The (constant) horizontal velocity component.

   (d) The horizontal distance between landing and launch points, using answers to parts (b) and (c).

5 A 10 N force acts on a mass for 5 seconds. What change in momentum does it produce? If the mass was 25 kg, and initially stationary, how fast will it be moving after 5 seconds?

6 A box of mass 30 kg sits on a rough floor and is pulled slowly sideways at constant velocity by a constant force of 70 N. Why is it not accelerating?
The force is increased to 130 N and the box does now accelerate.

   (a) What is the acceleration?

   (b) How long will it take to travel 4.0 m?

7 A satellite is in circular orbit around the Earth at a radius of 6670 km at which height the acceleration of free fall is 9.0 m s^{-2}.

   (a) Explain in your own words why it must be accelerating and in which direction the acceleration has to be.

   (b) How fast is it moving?

   (c) What is its angular velocity?

   (d) How long does it take to orbit the Earth once?

8 A mass of 2 kg travelling at 12 m s^{-1} collides with and sticks to a stationary mass of 6 kg. What is the combined velocity?

9 A ballistic pendulum is an arrangement for measuring the speed of a bullet whereby the bullet is fired into a much larger hanging mass, embedding itself. The two swing together to a measured height which relates to the speed of the bullet. The theory involves the law of conservation of momentum for the impact, followed by the law of conservation of energy for the conversion of kinetic into gravitational potential energy.
In a laboratory version the bullet of mass 5.0×10^{-4} kg is fired into a suspended mass of 0.05 kg at a velocity of 90 m s^{-1}.

   (a) Apply the law of conservation of momentum to find the velocity with which the block-bullet combination moves off just after the impact.

(b) Find the vertical height *h* risen by the combination as it swings upwards in a circular arc on its suspension. You should do this by considering the kinetic energy at the bottom converting into potential energy at the top of the swing.

10

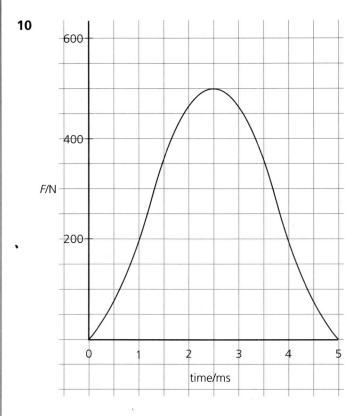

The graph shows how the force which a racket exerts on a ball varies with time during the impact. The ball was initially stationary and had a mass of 0.04 kg.

(a) Estimate from the graph the impulse on the ball and hence its change in momentum.
(b) Use this value to find the velocity with which it leaves the racket.
(c) What is the maximum acceleration of the ball?

11 Water squirts out of a pipe of cross-sectional area 0.01 m² at a speed of 15 m s⁻¹. The density of water is 1000 kg m⁻³. Calculate:

(a) The volume flow rate in m³ s⁻¹.
(b) The momentum given to the water per second.
(c) The force exerted by the jet if it hits a vertical wall while moving horizontally at the speed of 15 m s⁻¹.

12 In an experiment on collisions on an air track a vehicle of mass 0.2 kg moving at 1.2 m s⁻¹ catches up a vehicle of mass 0.1 kg moving at 0.8 m s⁻¹ and joins on to it.

(a) What is the velocity of the combination after impact?
(b) What is the total kinetic energy after collision?
(c) What was the total initial kinetic energy?
(d) What percentage of the original kinetic energy has been lost?
(e) How would you describe the collision?

13 A builder is carrying a horizontal ladder 4 m long on his shoulder. Its weight is 250 N and is uniformly distributed so that its centre of gravity is at the middle. The length projecting forwards from his shoulder is 1 m and his hand pulls down on the ladder at a point 0.75 m in front of his shoulder.

(a) Draw a diagram of the arrangement, marking in all the forces.
(b) Take moments about a suitable point to find the force with which he is pulling down on the ladder.
(c) What force does his shoulder experience?

14 What is the kinetic energy of:

(a) A 100 tonne train moving at 20 m s^{-1} (1 tonne = 1000 kg).
(b) A 10 gramme bullet fired at 200 m s^{-1}?

If the bullet were fired vertically how high would it go in the absence of air resistance?

15 A motorbike with rider has a kinetic energy of 75 kJ. The brakes produce a retarding force of 900 N. What is the stopping distance?

16 A catapult can store 30 J of elastic potential energy. How high can it project a stone of mass 5 grammes?

17 In a hydroelectric power scheme water falls through a vertical height of 50 m. What mass flow rate is required to generate 600 MW, assuming 100% efficiency of conversion?

18 What power does a train motor deliver if it pulls with a force of 5×10^4 N at a steady speed of 30 m s^{-1}?

If the power is proportional to the cube of the velocity, how much power is needed to drive the train at 40 m s^{-1}?

Chapter 2 Thermal physics and states of matter

Where necessary use the following values:
SHC of water = 4200 J kg^{-1} K^{-1}; SLH of vaporisation of water = 2.3×10^6 J kg^{-1};
SLH of fusion of ice = 3.3×10^5 J kg^{-1}; $R = 8.31$ J mol^{-1} K^{-1};
$N_A = 6.0 \times 10^{23}$ mol^{-1}; $g = 9.8$ N kg^{-1}

1 A centigrade mercury-in-glass thermometer scale is constructed by measuring the lengths at the lower and upper fixed points which are 13.2 mm and 41.0 mm respectively. The length at some intermediate temperature is 22.4 mm.
A platinum resistance thermometer is also calibrated: the lower and upper fixed point resistances are 4.86 Ω and 6.75 Ω. The resistance was then measured at the same intermediate temperature as with the mercury thermometer and found to be 5.47 Ω. Calculate the values of the same temperature on the two scales.

2 An electric kettle is rated at 2.1 kW and has a thermal capacity of 390 J K^{-1}. How long will it take to bring 1.2 kg of water from a temperature of 16 °C to boiling point?
The thermal cut-out fails and it continues to boil for 10 minutes. What mass of water will have been boiled away?

3 A night storage heater contains 60 kg of thermal bricks with a SHC of 800 J kg^{-1} K^{-1}. They are heated overnight from 15 °C to 40 °C and cool down the next day over a period of 5 hours.
(a) How much energy have they stored?
(b) What is the average power lost as they cool down?

4 An electric shower heater is rated at 8 kW and heats water which flows over it. The inlet water temperature is 15°C. The maximum permitted water flow rate is 0.1 kg s^{-1}.
(a) Why is this the *maximum* rate allowed?
(b) What is the shower temperature at this flow rate?

5 A car has a mass of 1000 kg and travels at 30 ms^{-1}. It has 4 disc brakes each of mass 5.0 kg and made of material of SHC 450 J kg^{-1} K^{-1}. If all the kinetic energy of the car is dissipated in the brakes what is their rise in temperature?

6 An ideal heat engine extracts heat from a hot reservoir at 1100 K and rejects heat to a cold reservoir at 300 K, in the process doing mechanical work. If its output mechanical power is 100 kW at what rate must the cooling system remove heat from the cold reservoir?

7 The ice on a pond is 5 mm thick and has a temperature difference across it of 0.5 K. The thermal conductivity of ice is 2.3 W m^{-1} K^{-1}.

(a) At whate rate per m^2 is heat being extracted from the water below the ice?

(b) The water just below the ice is at its freezing point. At what rate in kg s^{-1} is new ice being formed on the lower face of the existing layer for each 1 m^2 of surface?

(c) The density of ice is 920 kg m^{-3}. At what rate is the thickness increasing in mm per hour?

(d) Why would it not in fact increase by this amount in 1 hour?

8 The U-value for a window is 4.5 Ω m^{-2} K^{-1}. What is the rate of heat loss through a window of area 1.7 m^2 if the outside temperature is 5°C and the inside 20°C?

9 An atom has a diameter of approximately 1.0 × 10^{-10} m. Treating it as a cube:

(a) What is its effective volume?

(b) How many atoms are there in 10 cm^3?

(c) How many moles does this represent?

(d) If the density is 2500 kg m^{-3} what is the molar mass?

10 A cylinder has a volume of 1.5 m^3. Nitrogen at 300 K has been pumped in to produce a pressure of 2.0 MPa. The molar mass of nitrogen is 0.028 kg mol^{-1}.

(a) How many moles of nitrogen are in the cylinder?

(b) What mass is this?

(c) What is the density of nitrogen under these conditions?

11 A fixed mass of gas is at a pressure of 0.2 MPa and a temperature of 300 K. The pressure is increased to 0.5 MPa and the temperature is also adjusted so that the density of the gas has increased by a factor of 4. What is the new temperature?

(Hint: Rearrange the ideal gas equation to have density rather than volume as the third variable.)

12 The molecules of a monatomic gas have a mean kinetic energy of 1.0 × 10^{-20} J.

(a) What is the gas temperature?

(b) If the molar mass is 4.0 × 10^{-3} kg mol-1 what is the rms speed of the molecules?

(c) What is the thermal kinetic energy of 1 mole?

(d) If all of this energy could be converted into gravitational potential energy, what extra height would the mass be at?

13 The density of argon is 3.2 kg m^{-3} at a pressure of 0.2 MPa.

(a) What is the rms speed of the molecules?

(b) If the mass of 1 molecule is 6.6 × 10^{-26} kg what is the mean KE per molecule?

(c) What is the temperature of the argon gas?

(d) If the pressure were doubled, keeping the temperature the same, what would happen to the molecular speed?

14 In a laboratory test on steel using a tensometer a narrow steel rod of diameter 5.0 mm was stretched by a force of 2 kN. The initial length of the rod was 0.40 m. Young's modulus for steel is 1.9 × 10^{11} Pa. Calculate:

(a) The stress in the steel, and hence the strain (assuming Hooke's law to hold over this range).

(b) The extension of the rod.

15 The lift cage in a coal mine is supported by 1 km of steel cable of effective diameter 20 mm. How much will the cable stretch by if 12 miners of average mass 80 kg enter it at the bottom of the shaft? (Young's modulus for steel is 190 GPa).

16 A bungee jump rope stretches 3.5 m when a jumper of mass 80 kg hangs in equilibrium from it.

(a) What is the stiffness of the rope?

During a jump the person drops a maximum height of 50 m (when the rope is at its fullest stretch and the jumper is momentarily at rest).

(b) Assuming 100% conversion of gravitational potential energy into elastic energy, and that the rope obeys Hooke's law, how much does the rope stretch by during the jump?

(c) What was the original length of the rope?

(d) How fast was the person falling when the rope just started to go under tension?

Where necessary use the following values:
$e = 1.6 \times 10^{-19}$ C; $N_A = 6.0 \times 10^{23}$ mol^{-1}

1 A current of 10 mA flows through a copper wire of diameter 0.56 mm. There are 1.0×10^{29} conduction electrons per m^3 of copper. What is their drift velocity?

2 Copper has a density of 8900 kg m^{-3} and a molar mass of 0.064 kg mol^{-1}. Assuming that each copper atom contributes one conduction electron, use these data to confirm the electron density value used in Q. 1

3 Semiconductors have much lower electron densities than metals: for germanium it is 6.0×10^{20} m^{-3}. A slice of germanium of the kind used in a Hall probe carries a current of 5.0 mA through a rectangular cross-section of width 5.0 mm and thickness 1.0 mm. What is the electron drift velocity?

4 A power supply of emf 12 V delivers a current of 2.3 A into a small heater. It flows for 10 minutes.
   (a) How much charge passes?
   (b) How much energy is converted by the supply?

5 An electric motor lifts loads through a height of 20 m at a rate of 10 tonne per hour (1 tonne is 1000 kg).
   (a) What is the power output of the motor?
   (b) The motor is supplied by 400 V and draws 1.8 A. What is the efficiency of power conversion?

6 A 3 Ω and a 6 Ω resistor are connected together in series and then in parallel across a 12 V supply. Calculate the power dissipated in each resistor in each case.

7 An electric train of mass 150 tonne is supplied with power at 25 kV. It can reach a speed of 25 m s^{-1} in 40 seconds.
   (a) What is the average rate of increase of kinetic energy?
   (b) Assuming 100% efficiency what is the average current drawn?

8 An electrical data table shows the resistance per metre of nichrome wire of diameter 0.25 mm to be 22.0 Ω m^{-1}. What is the resistivity of nichrome? What length of this wire is needed to make a 24 W heating coil working off a supply of 12 V?

9 A wire has a resistance of 12 Ω. If the length is doubled and the diameter halved what is the new resistance?

10 5 metres of wire of diameter 0.2 mm with a resistivity of 8×10^{-8} Ω m are connected across a 10 V supply. Find the values of:
   (a) resistance
   (b) current
   (c) power dissipated.

11 A battery of emf 12 V is connected to a 10 Ω resistor and a high resistance voltmeter connected across its terminals reads 9 V. Find the values of:
   (a) current
   (b) internal resistance of the battery
   (c) rate of energy conversion in the battery
   (d) power dissipated in the 10 Ω load.

12 In the circuit below the three resistors each have 4 Ω and the battery has emf 6 V and negligible internal resistance.

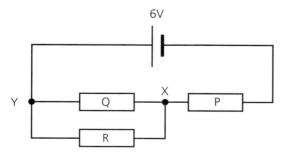

What is the p.d. between points X and Y? How would the value change if each resistance was doubled?

13 In the circuit below the LDR and the fixed 10 kΩ resistor are connected as a potential divider across 6 V. In the dark the LDR has a resistance of 0.5 MΩ and in the light 800 Ω.

(a) Describe qualitatively what happens to the potential at the junction point X as the light intensity on the LDR is gradually increased.

(b) Calculate the dark and light values of the potential.

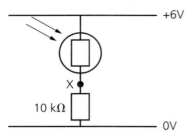

14 In order to limit the current through a light emitting diode (LED) to 10 mA it should be used with a series resistance of 220 Ω when connected to 6 V. Find the values of:

(a) the p.d. across the protecting resistor

(b) the p.d. across the LED

(c) the resistance of the LED for this current.

Why could the LED not just be labelled with this resistance?

15 Apply Kirchhoff's Laws to the circuit below to find the currents in each of the three resistors. Since there are three unknown quantities you will need to construct three equations: one can come from applying the 1st Law to one of the junctions and the other two will come from applying the 2nd Law to two of the three loops in the circuit.

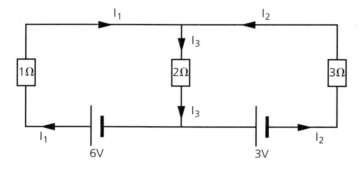

16 The graph shows the current-voltage characteristics for two lamps.

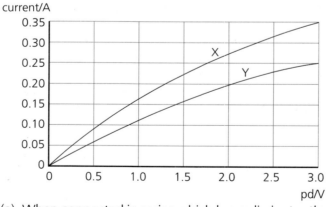

(a) When connected in series which lamp dissipates the larger power?

(b) When connected in parallel which lamp dissipates the larger power?

(c) When connected in series what p.d. is needed to deliver 0.20 A?

(d) How much current would flow through a 2.0 V supply when connected to X and Y in parallel?

1 The displacement x metres of an object moving with SHM is given by the equation
$x = 0.2\cos(10t)$
where t is the time in seconds from the start and the angle $10t$ is in radians.
What are the values of:
(a) the amplitude
(b) the frequency?
If the mass of the object is 0.2 kg what is:
(c) the stiffness of the elastic force producing the motion
(d) the total energy?
After 0.4 seconds how large is:
(e) the displacement
(f) the acceleration?

2 The level of sea water rises and falls with SHM according to the state of the tide and has a period of 12.5 hours. The depth of water in a harbour varies from 5 metres to 15 metres and is at its lowest at midday.
(a) Sketch a graph of water depth against time of day (pm), showing particularly the amplitude and mean sea level.
(b) Show that the depth h metres follows the relation $h = 10 - 5\cos(0.5t)$ where t is the time in hours since midday and the angle $0.5t$ is in radians.
A boat entering the harbour needs a minimum depth of 7.5 m.
(c) Show on your graph the approximate earliest time in the afternoon at which the boat can enter the harbour.
(d) Calculate the exact time, from the SHM equation in (b).

3 A particle of mass 0.005 kg oscillates with SHM with amplitude 5.0 mm. The maximum force acting on it is 4.8 N. What are the values of:
(a) maximum acceleration
(b) frequency
(c) maximum velocity?

4 An ultrasonic oscillator has a frequency of 40 kHz and amplitude of 2×10^{-6} m. What are the values of maximum velocity and acceleration? Does the acceleration value have any implications for the mechanical design of the oscillator?

5 A spring is pulled with 20 N force and it stretches by 4.0 cm. It is hung vertically, a 4.0 kg mass is attached and it is pulled down a further 3.0 cm and released.
(a) What is the spring stiffness?
(b) What is the amplitude of the oscillations?
(c) What is the mean extension of the spring during the oscillations?
(d) What is the period of oscillation?
(e) What is the maximum kinetic energy of the 4.0 kg oscillating mass?

6 This question is to give you practice in some sensible estimating.
An elephant walks across a simple plank bridge. By thinking about possible values for its mass and the deflection it might produce when standing in the middle of the bridge, estimate its possible frequency of bouncing when on the bridge. Should a herd of elephants be encouraged to break step when crossing the bridge to avoid resonance damage?

7 An astronaut's body mass is monitored by placing her in a seat fixed to a stiff flexible beam and observing changes in the frequency of vibrations. The pre-flight calibration with a mass of 64.21 kg gave a frequency of 0.4814 Hz. After some time in orbit the frequency was measured to be 0.4709 Hz. What was her new mass? (Be careful with significant figures here – more than are usually required!)

8 The stiffness of a car's suspension at one wheel was estimated by observing that a 70 kg person sitting on the body over the wheel lowered it by 5 cm. The wheel's mass was estimated to be 10 kg. What do these values give for:
(a) the stiffness (in N m^{-1})
(b) the natural frequency of oscillation of the wheel suspension spring system?
The car is driven over regularly spaced bumps 3.5 metre apart. What speed should be avoided if the wheel is not to resonate?

It is possible to have resonance of the whole car body if driven over the same bumps. If the car's mass is 1000 kg what speed should be avoided for the same reason? (Think about the car bumping up and down on 4 springs.)

Where necessary use the following values:
velocity of sound in air (unless otherwise stated) = 330 m s^{-1}
velocity of electromagnetic waves in a vacuum c = 3.0 × 10^8 m s^{-1}

1 The diagram shows a transverse wave pulse (in reality the edges would be more rounded) travelling at a speed of 2.0 m s^{-1} just arriving at point A. B is a point 4.0 m further on from A. The pulse height is 0.5 m.
   (a) Sketch a graph showing how the vertical displacement of A varies with time, taking $t = 0$ as the instant when the pulse arrives at A.
   (b) On the same axes sketch also the motion of B.
   (c) Comment on the shape of the displacement–time graphs compared with the original pulse shape of displacement against position at a fixed time instant.
   (d) What is the actual velocity of A as it moves upwards?

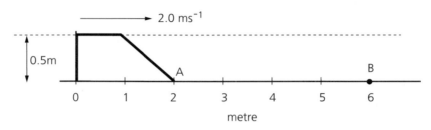

2 The velocity of longitudinal waves in a solid is given by the expression
   $\sqrt{\{\text{Young's modulus/density}\}}$
   Check that this does have the units of velocity.
   What is the Young's modulus for aluminium if its density is 2700 kg m^{-3} and the velocity of sound in it is 5100 ms^{-1}?

3 The speed of sound in a gas is given by the expression
   v = constant × $\sqrt{\{p/\rho\}}$
   where the constant has no units (a 'dimensionless constant'), p is the pressure and ρ the density.
   (a) Check that this expression has the units of velocity.
   (b) Show that for an ideal gas $v = k\sqrt{T}$ where T is the kelvin temperature and k is a constant for a particular gas.
   (c) If the speed of sound in air is 340 ms^{-1} at 300 K, what is it at an altitude where the temperature has dropped to 200 K?

4 An optical fibre has a core of refractive index 1.47 and an outer sheath of refractive index 1.45 .
   (a) What is the critical angle for this interface?
   (b) Sketch what happens to a ray in the core which strikes the boundary at angles of incidence:
        (i) less than the critical angle
        (ii) equal to the critical angle
        (iii) greater than it.
   (c) Light which travels the zig-zag path down the fibre will travel a longer distance than light which goes straight down. Calculate the *maximum* value of this extra distance for each kilometre length of fibre.

5 In a Young's double slit interference experiment with monochromatic light a pair of slits 0.50 mm apart were used and the interference pattern was viewed at a distance of 750 mm from them. 11 fringes were counted (think carefully what this means!) in a length of 8.1 mm.
   (a) What was the wavelength of light used?
   (b) You have control over slit separation, slit-observer distance and the colour of the light.

How would you change these one at a time to make the interference fringe pattern broader?

6 In the diagram T is a wave transmitter, R a receiver and there is a reflector parallel to TR and some distance from it. Waves can reach R either directly or by the longer reflected route.

(a) If the reflector is slowly moved along the line perpendicular to TR the response of R is a sequence of maxima and minima. Explain how this happens.

(b) The first position shown produces a maximum at R and the second one a minimum. What are some possible values for the wavelength? What else would you have to know to be certain?

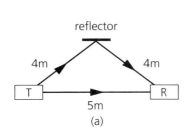

(a)

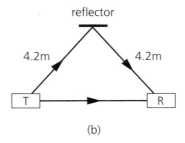

(b)

7 The velocity of transverse waves down a taut string is given by the expression
$$\sqrt{\{\text{tension/mass per unit length}\}}$$

The vibrating length of a stringed instrument is 0.40 m and is tightened to a tension of 18 N. Its mass per unit length is 4.3×10^{-4} kg m^{-1}.

(a) What is the wavelength of the standing wave on the string in its fundamental mode?

(b) What is the fundamental frequency?

(c) What is the wavelength of the *sound in air* of the second harmonic?

8 Two long wave radio transmitters are set up at a separation of about 200 km, broadcasting at a frequency of 200 kHz. A reasonably straight section of motorway runs down the line joining the transmitters. Describe the pattern of radio reception in a car travelling along the motorway.

9 A tuning fork tuned to 384 Hz is used to produce resonance in a closed pipe in air. What are the first two lengths of the air column which will resonate at this frequency?

10 A diffraction grating has 600 lines mm^{-1} and is used with red light of wavelength 6.5×10^{-7} m and blue light of 4.0×10^{-7} m .

(a) Calculate the grating spacing s in metres.

(b) Find the first order diffracting angles for both colours.

(c) Repeat for the second order.

(d) You should find that the angle between the two colours is bigger for the second order and it is generally true that the angular separation between a pair of wavelengths increases as the order increases. Sodium yellow light has a pair of wavelengths very close to each other and in order to see them clearly as high an order of diffraction as possible is needed. Each wavelength is very close to 5.9×10^{-7} m: can they be viewed with this grating in the 3rd order?

Chapter 6
Gravitational and electric fields

Where necessary use the following values
$G = 6.67 \times 10^{-11}$ N m^2 kg^{-2}; g on Earth's surface $= 9.8$ N kg^{-1}
$\varepsilon_0 = 8.85 \times 10^{-12}$ F m^{-1}; $e = 1.6 \times 10^{-19}$ C; $R_e = 6.4 \times 10^6$ m

1 A light sphere carries a charge of 30 nC and hangs between two parallel metal plates 8 cm apart connected to a high voltage source. The sphere has a mass of 0.5 gramme and its suspension makes an angle of 5° with the vertical.

(a) Draw a diagram marking *three* forces acting on the ball and either by scale drawing (not very accurate here unless the triangle of forces is drawn on a full sheet of paper) or by resolving horizontally and vertically find the size of the horizontal force on the ball.

(b) How big is the p.d. across the plates?

2 Two large metal parallel plates are 15 mm apart and have 5 kV pd across them.
 (a) What is the potential gradient in volt/metre in the gap?
 (b) What is another way of stating your answer to (a)?
 (c) What force acts on a speck of dust carrying 10^8 surplus electrons when it is the middle of the space between the plates?
 (d) Describe roughly how you would expect the force to vary as the speck moves around in the 3–dimensional space.

3 In a version of Millikan's oil drop experiment the drop, of mass 1.5×10^{-15} kg, is balanced in the space between the horizontal plates when the p.d. between them is 300 V. They are 10 mm apart. How many surplus electrons are there on the drop? In which direction is the E-field between the plates?

4 A parallel plate capacitor is made from plates 0.25 m square and separated by a gap of 2.5 mm. A 12 V supply is connected across them. What is the:
 (a) capacitance
 (b) charge on one plate
 (c) charge density σ
 (d) value of σ/ε_0
 (e) electric field in the gap?
 It is found that when the gap is doubled (still with 12 V connected) and the new charge measured it has fallen by significantly less than the predicted 50%. Suggest a reason for this.

5 One way of accurately measuring the small charges such as those in Q.4(b) above is to use a reed switch, where the capacitor is charged and discharged very rapidly at a known frequency and the mean discharge current is measured. A 0.2 nF capacitor is charged and discharged by an electronic reed switch at a frequency of 50 kHz to a p.d. of 12 volt. What average current flows in the discharge circuit?

6 In a design for a capacitor driven electric motor for a vehicle it is proposed to use four 0.1 F capacitors in parallel. A design criterion is that they should be able to store enough energy to give an 800 kg vehicle a velocity of 20 ms^{-1}. What charging voltage would be required?

7 The circuit below is set up and the charge on the 10 μF capacitor was measured at regular intervals after opening the switch. The charge–time graph is shown and you are asked to make measurements from it.

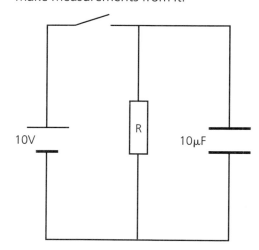

10V   R   10μF

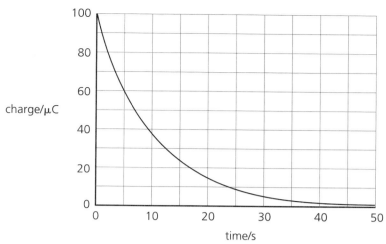

(a) Find the gradients at t = 0, 10 and 20 s. What do these gradient values represent (think of the units)? Do the values suggest the decay of charge is exponential? Check your answer by looking at the values of charge itself at these times.

(b) From the gradient at $t = 0$ and the battery voltage calculate the value of R.

(c) The time constant (=RC) is the time taken for the charge to drop to 37% of its initial value. From the graph estimate the time constant and hence check your value of R found in (b).

(d) If the capacitor was halved.
   (i) state *2 ways* in which the graph would change
   (ii) what happens to the initial current?

8 In a linear accelerator protons are accelerated over a distance of 1 kilometre almost to the velocity of light, to a kinetic energy of 20 GeV.
   (a) Express this energy in joules.
   (b) What potential difference could produce this energy?
   (c) What E-field would produce the necessary acceleration?
   (d) If you calculate the velocity for this kinetic energy in the normal way why do you get an absurd value?

9 If the force of repulsion between two small spherical charges each of 5 nC is 10 μN how far apart are they?

10 (a) Using the surface value of the Earth's gravitational field and the radius of the Earth, calculate the mass of the Earth.
   (b) Now find the mean density of the Earth.
   (c) Rocks picked up or excavated near the surface of the Earth have a density much less than the value found in (b). Suggest some reasons for this.
   (d) Put your calculations in (a) and (b) into algebraic form to finish up with an expression for the surface gravitational field strength in terms of density and radius. Use it to decide whether large astronomical bodies of about the same density as smaller ones have larger or smaller surface gravitational fields.

11 What is the gravitational field strength on the surface of a neutron star of mass equal to that of the Sun (2.0×10^{30} kg) and of radius 1.5 km?

12 (a) What is the angular velocity in radian/second of the Earth about the Sun?
   (b) The mean distance of the Earth from the Sun is 1.5×10^{11} m. Apply Newton's Second Law to the orbital motion of the Earth around the Sun and using the numerical values for angular velocity and distance calculate the mass of the Sun.

13 (a) Calculate the gravitational potentials at distances of 10×10^6 and 20×10^6 metres from the centre of the Earth.
   (b) A space craft of mass 800 kg moves from the equipotential at 10 Mm to that at 20 Mm. What is the change ΔE_g in its gravitational potential energy ?
   (c) It later returns and crosses the outer equipotential at a velocity of 3 km s^{-1}. How fast will it be moving when it crosses the inner one (at 10 Mm)?

14 Following the fission of uranium, possible products are the nuclei of Rb (37 protons) and Cs (55 protons). What is the electrostatic potential energy of this pair of nuclei as they start to fly apart from a separation of 1×10^{-14} m?

Where necessary use the following values:
$\mu_0 = 4\pi \times 10^{-7}$ T m A^{-1}; $e = 1.6 \times 10^{-19}$ C; $m_e = 9.1 \times 10^{-31}$ kg

1 A current of 5.0 A flows in a straight wire in a uniform B-field of 2.0 mT. What is the force acting on the wire per unit length if the wire is:
(a) perpendicular to the field
(b) at 60° to the field
(c) parallel to the field?

2 A particle of charge q and mass m moves at right angles to a magnetic field of strength B. Show that it moves in a circle with angular velocity ω given by
$\omega = Bq/m$
In a laboratory experiment with a fine beam tube an electron moves in a field of 1.0mT. How many revolutions per second does it make in the tube?

3 An electron is bent into a circular path of radius 120 mm by a B-field of 1.5 T. What is its momentum?

4 A straight wire of length 0.75 m moves at 12 m s^{-1} at right angles to a B-field of 25 mT. Its resistance is 5.0 Ω. What current flows if its ends are joined by wires of negligible resistance?
Why would you have to be careful about the connecting arrangement in order to produce the current?

5 A coil of cross-sectional area 10 cm^2 has 4000 turns. Its plane is perpendicular to a B-field of 0.1 T and it is then rotated so that the plane ends up at 30° to the field. This rotation takes 0.5 seconds.
(a) What is the initial flux linkage through the coil?
(b) What is the final flux linkage?
(c) How large is the mean emf induced in the coil?

6 In an experiment to demonstrate mutual induction one solenoid is placed inside another. The current in the outer one is increased steadily from zero to 2.5 A in 2.5 seconds and a voltmeter connected to the inner shows a constant emf of 7.5 mV.
(a) What is the rate of change of current in the outer solenoid?
(b) Hence calculate the coefficient of mutual induction between the coils.
(c) With the current in the outer coil still at 2.5 A, how would you remove the inner one so as to produce a nearly steady emf of the same value as before? What would its direction be? How could you produce 10 mV in the same way?

7 A series circuit consists of a 12 V battery, a pure inductor of 2.5 H, a resistor of 9.0 Ω and a switch. Some time after the switch is closed the current is I and is changing at a rate dI/dt.
(a) Draw the circuit and mark on it the voltages across the inductor V_L and resistor V_R in terms of dI/dt and I.
(b) Write down a relation between these two voltages and the battery voltage
(c) At the instant the switch is closed what can you say about the values of I and dI/dt?
(d) What happens to these values as time increases?
(e) Sketch a graph of current against time, putting as much numerical information on as possible.

8 It is important for a particular experiment that it is carried out in zero horizontal B-field. It is proposed to conduct the experiment inside a large solenoid which will carry just the right current to cancel the Earth's horizontal component of 1.8×10^{-5} T. If the solenoid is 0.5 m long with 300 turns how large should this current be?
If by mistake the solenoid is lined up magnetic east–west what can you say about the resultant B-field well inside?

9 Because of accessibility problems it was decided to carry out the above experiment at the centre of a flat circular coil carrying 100 turns with a mean radius of 10 cm. What would be the current now?

10 How large is the B-field at a distance of 1.5 cm from a long straight wire carrying a current of 10 A?
A second long wire carrying the same current in the opposite direction is placed parallel to the first and 1.5 cm from it.

(a) Draw a diagram (or two) showing the currents, and fields produced by the currents, and use it to show that the wires must repel each other.

(b) Find the value of the force per unit length on the second wire.

11 (a) A circuit used for charging a capacitor works from a 12 V rms AC supply but charges up the capacitor to *twice* the peak voltage. If the capacitance is 50 μF, what is the charge on one plate?

(b) An underground power cable carries 100 kV rms. What should be the minimum voltage the insulation should be designed to withstand?

12 A power supply delivers a sinusoidal voltage of peak value 15 V to a resistor of resistance 5 Ω.

(a) Sketch 3 separate graphs, but using the same time axis, to show how the following quantities vary over at least two cycles:
  (i) voltage
  (ii) current through the resistor
  (iii) power dissipated in the resistor.

(b) State the values of:
  (i) rms current
  (ii) mean power.

13 An oscilloscope is used to measure an alternating current by displaying and measuring the pd across a fixed resistor of 3.3 Ω carrying the current. If the *peak-to-peak* height of the trace is 5.2 cm and the Y sensitivity is 2 mV/cm, what is the value of the rms current?

14 An rms current of 13 A is drawn from the 50 Hz AC mains supply. If the current can be represented by the relation:

$I = I_0\sin\omega t$

what are the values of I_0 and ω?

How large is the current 2.5 ms after passing through zero?

15 A power station delivers 2200 MW at a generating voltage of 11 kV.

(a) How much current flows?

(b) If the cables have an overall resistance of 5.0×10^{-5} Ω m^{-1} what is the power loss due to heating of the cable per kilometre length?

(c) If the voltage is transformed up to 400 kV how do these values change?

16 This question is about AC through capacitors and inductors – check it is in your syllabus before trying it!

An LCR series circuit consists of a 1.0 μF capacitor in series with a pure inductor L henry and 10 Ω. An alternating p.d. at a frequency f is across the combination.

(a) What is the reactance of the capacitor at 50 Hz?

(b) If the circuit has zero reactance at 100 Hz what is the value of L?

(c) For this value of L, sketch how the current amplitude will vary over the frequency range 90–110 Hz (accurate calculation not required).

(d) If the rms voltage is 5 V what is the rms current at 100 Hz?

Chapter 8
Nuclear physics

Where necessary use the following values

$e = 1.6 \times 10^{-19}$ C; $c = 3.0 \times 10^{8}$ m s^{-1}; 1u $= 1.66 \times 10^{-27}$ kg

$1/4\pi\varepsilon_0 = 9.0 \times 10^{9}$ F^{-1} m

1 The effective radius of a nucleus is given approximately by the formula

$r = 1.4 \times 10^{-15}$ A$^{1/3}$ metre

where A is the nucleon number.

(a) Using this expression calculate the radius of a copper nucleus (A = 63) and hence its volume, assumed spherical.

(b) Copper has a molar mass of 0.063 kg mol^{-1} and a density of 8900 kg m^{-3}. Find the effective volume taken up by 1 copper atom and calculate the ratio (volume of atom/volume of nucleus).

(c) What therefore is the approximate mean density of the nucleus?

(d) What is the radius of a neutron star of this density which has the mass of the Sun, 2.0×10^{30} kg?

2 An alpha particle (charge +2e and mass 4u) is moving directly towards a stationary

uranium nucleus (charge $+92e$ and mass 235u) at a speed of 1.8×10^7 m s^{-1} and recoils straight back along its original path.
(a) Discuss the energy changes during this process, whether or not you need to consider the motion of the uranium nucleus and whether you would describe the interaction as elastic or inelastic.
(b) Use your energy considerations to find the distance of closest approach.
(c) How would the value change for a proton of the same kinetic energy?

3 The nuclide $^{235}_{92}$U is the start of a naturally occurring chain of radioactive transformations which undergoes 11 changes (4 β and 7 α) to finish with $^{207}_{82}$Pb. Use a table or chart of nuclides to track through the intermediate stages.

4 In an experiment to measure the half-life of radon the gas was introduced into an ionisation chamber and the ionisation current measured. The current is proportional to the number of alpha particles in the chamber and hence to the activity of radon. The following readings were taken at 30 second intervals:

time/s	0	30	60	90	120	150	180
current/nA	201	142	101	71	56	43	38

The current settled down to a steady reading of 21 nA after a long time, attributable to background radiation.
(a) Plot a graph (corrected for background) of current against time.
(b) Comment on any of its features.
(c) Measure the half-life using 3 different pairs of values and find the mean.
(d) Also plot ln(corrected current) against time. The gradient of the best fit straight line gives the negative value of the decay constant λ. Find λ and use it to find the best value for half-life: compare with the value from (c).

5 A radioactive sample is known to have a half-life of 85 seconds and at a particular instant its activity is 7.6×10^3 Bq. How many nuclei of this isotope are present in the sample?

6 The radioactive isotope ^{60}Co is being considered for a small self-contained power supply. Its half-life is 5.3 years and each decay produces radiation with a total energy of 2.6 MeV.
(a) If the initial power is required to be 20 mW how many decays per second does this represent?
(b) Find the decay constant λ and hence the initial number of nuclei of the cobalt isotope.
(c) Taking the mass of the atom as 60u, what is the approximate mass of the sample?
(d) The minimum power requirement for the application is between 2 and 3 mW. What is the effective lifetime of the supply?

7 The decay of potassium into argon by the following process
$^{40}_{19}$K $\rightarrow$ $^{40}_{18}$A $+$ $^{0}_{1}\beta$ (note positron decay)
can be used for very long time scale geological dating.
A sample of rock contains 0.8 μg of the potassium isotope and 5.6 μg of argon. The activity is 0.21 Bq.
(a) Show that the number of potassium nuclei in the sample is approximately 1.2×10^{16}.
(b) Hence show that the decay constant is 1.8×10^{-17} s^{-1} and use this value to find the half-life in years.
(c) What is the age of the rock, assuming it started with no argon and that all the argon produced remained trapped in the rock.

8 The following hydrogen fusion reaction (using the hydrogen isotope *deuterium* – the nucleus is called a *deuteron*) is one being considered for fusion power generation:
$^{2}_{1}$H $+$ $^{2}_{1}$H $\longrightarrow$ $^{3}_{2}$He $+$ $^{1}_{0}$n
The masses are: deuteron 2.01410 u; ^{3}He 3.01603 u; neutron 1.00866 u .
(a) Find the mass loss in kg per fusion.
(b) Hence find the energy released in:

(i) MeV per fusion

(ii) J per kg of hydrogen.

9 One of the many possible uranium fission reactions is:

$$^{235}_{92}U + ^{1}_{0}n \longrightarrow ^{144}_{56}Ba + ^{90}_{36}Kr + 2\,^{1}_{0}n$$

The nuclear masses of these isotopes are:

U 235.044 u; Ba 143.922 u; Kr 89.919 u; neutron as in Q.8.

Calculate the energy released in this reaction:

(a) in J per fission

(b) in MeV per fission

(c) in J per kg of uranium.

Where necessary use the following values:

$h = 6.6 \times 10^{-34}$ Js; $c = 3.0 \times 10^{8}$ m s^{-1}; $e = 1.6 \times 10^{-19}$ C;

$k = 1.4 \times 10^{-23}$ J K^{-1}; $m_e = 9.1 \times 10^{-31}$ kg

1 (a) Find the energy of a photon emitted by a helium-neon laser at a wavelength of 633 nm:

(i) in J

(ii) in eV.

(b) If the laser power is 0.70 mW, how many photons are emitted per second?

(c) If the beam diameter is 2.0 mm, how many photons per m^3 are there in the beam – the photon density (remember they travel at the speed of light!)?

2 (a) If a metal surface has a work function of 2.5 eV what is the longest wavelength of light which will eject photoelectrons?

(b) If this surface is radiated with ultraviolet light of wavelength 2.8×10^{-7} m

(i) what is the maximum velocity of the electrons?

(ii) what reverse voltage will just reduce the photoelectric current to zero (the stopping potential)?

3 The maximum kinetic energy of electrons emitted from a photocathode is 1.5 eV when light of frequency 7.0×10^{14} Hz is used. What is the minimum frequency needed for any emission at all? In which part of the electromagnetic spectrum is this light?

4 The diagram below shows the first 6 energy levels of the mercury atom. The energies are in eV.

```
6  ────────────────  −1.56
5  ────────────────  −2.67
4  ────────────────  −3.70

3  ────────────────  −4.94
2  ────────────────  −5.50




n = 1  ──────────────  −10.43
```

(a) The green line in the mercury spectrum has a wavelength of 5.46×10^{-7} m. What is the photon energy of this light in:

(i) J

(ii) eV.

(b) Which transition is responsible for this light?

(c) Mercury also emits in the ultraviolet. The longest UV wavelength is detected by a diffraction grating of spacing 600 lines per mm at an angle of 8.7° in the first order. Which transition is responsible for this UV light?

(d) Ultraviolet light of this wavelength is passed through cold mercury vapour. What happens to

(i) the mercury atoms in the vapour

(ii) the intensity of the transmitted light?

5 (a) The first excited state in the hydrogen atom is 10.2 eV above the ground state. What velocity should an electron have if it is to make a totally inelastic collision with a ground state hydrogen atom, raising it to the first excited state?

(b) What temperature should a gas of hydrogen atoms have so that their mean kinetic energy is sufficient for this excitation? Why will excitation (and even ionisation) occur at much lower temperatures than this?

6 Calculate the momentum and de Broglie wavelength for an electron and a proton if they each have a kinetic energy of 5.0 keV.

7 The wave-particle duality for atoms can be shown by the equivalent of the double slit interference experiment for light but using helium atoms. They are fired through a pair of slits only 1.0×10^{-5} m apart and are detected at a distance of 0.8 m away. A sensitive detector is moved parallel to the line of the slits and records maximum and minimum rates of arrival of helium atoms (equivalent to interference fringes) at intervals of 12 μm between maxima.

(a) What is the de Broglie wavelength for these atoms?

(b) How fast are they travelling if their mass is 6.6×10^{-27} kg ?

8 In a simple wave model of the hydrogen atom the electron is regarded as a standing de Broglie wave of wavelength twice the diameter of the atom (rather like a standing wave in an open pipe). The atom radius is 1×10^{-10} m.

(a) What momentum is associated with this wavelength?

(b) What is the kinetic energy (use $p^2/2m$)?

(c) For other wave modes for this size atom discuss whether these values represent maximum or minimum possible sizes.

(d) What electrostatic potential energy does the electron have?

(e) What therefore is the total energy? Comment on its sign.

(f) How would your answers to the above be changed if the electron was forced to be 10 times closer to the nucleus?

Answers

Answers have been quoted to the no. of significant figures in the question (as you should in an exam – you will be penalised if you quote too many). Intermediate calculations have however been carried out to a higher accuracy to avoid rounding errors so there may be small apparent discrepancies between individual answers.

1 (a) Small triangle is a quarter of the area of the larger one.
  (b) Each side of the smaller triangle is half that of the larger, making the area one quarter.

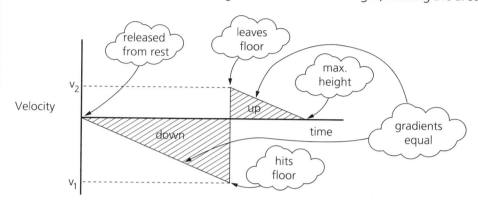

2 $\sqrt{(6^2 + 4.5^2)} = 7.5$ so the force is 7.5 kN
Angle to N is $\tan^{-1}(6/4.5) = 53°$

3 49 N

  (a) 49 N/cos30° = 57 N
  (b) $F = 28$ N
  (c) $F = 49$ N × tan20° = 18 N
    $F = 49$ N × tan 10° = 8.6 N

As the angle becomes smaller F becomes more nearly proportional to the angle (and hence sideways displacement).

4 (a) 30 m s^{-1} × sin60° = 26 m s^{-1}
  (b) use $v = u + at$ for the vertical motion; 2.7 s
  (c) 30 m s^{-1} × cos60° = 15 m s^{-1}
  (d) 15 m s^{-1} × 2.7 s = 40 m

5 50 N s or 50 kg m s^{-1}
  2 m s^{-1}

6 Opposed by an equal frictional force
  (a) (130 – 70)N/30 kg = 2.0 m s^{-2}
  (b) use $x = 1/2\,(at^2)$; 2.0 s

7 (a) Velocity is continually changing direction. The only force acting is gravity towards the centre of the Earth so this is the acceleration direction.
  (b) $v^2/r = g \Rightarrow v = 7750$ m s^{-1}
  (c) $v/r = 1.2 \times 10^{-3}$ rad s^{-1}
  (d) 5410 s = 1.50 h

8 3 m s^{-1}

9 (a) $v = (5 \times 10^{-4} \times 90) \text{ kg m s}^{-1}/(5 \times 10^{-4} + 0.05) \text{ kg} = 0.89 \text{ m s}^{-1}$
(b) $1/2(mv^2) = mgh \Rightarrow h = v^2/2g = 0.04 \text{ m}$

Note that as far as energy conservation is concerned, any sideways motion doesn't matter.

10 (a) The total area is needed: approximating to 2 triangles, area is
$(1/2 \times 500 \times 2.5 \times 10^{-3}) \text{ Ns} \times 2 = 1.25 \text{ N s or } 1.25 \text{ kg m s}^{-1}$
(b) $v = 1.25 \text{ kg m s}^{-1}/ 0.04 \text{ kg} = 30 \text{ m s}^{-1}$
(c) using the maximum force, accel $= 500 \text{ N}/0.04 \text{ kg} = 1.3 \times 10^4 \text{ m s}^{-2}$

11 (a) area × velocity $= 0.15 \text{ m}^3 \text{ s}^{-1}$
(b) $0.15 \text{ m}^3 \text{ s}^{-1} \times 1000 \text{ kg m}^{-3} \times 15 \text{ m s}^{-1} = 2250 \text{ kg m s}^{-2}$
(c) rate of destroying horizontal momentum $= 2250 \text{ N}$

12 (a) $\{(0.2 \times 1.2) + (0.1 \times 0.8)\} \text{ kg m s}^{-1} / (0.3 \text{ kg}) = 1.1 \text{ m s}^{-1}$
(b) 0.172 J
(c) 0.176 J
(d) 2.2%
(e) slightly inelastic

13 (a) forces on ladder: 250 N down at centre; upwards force P at shoulder; downwards force Q from hand
(b) taking moments about shoulder $Q \times 0.75 \text{ m} = 250 \text{ N} \times 1\text{m} \Rightarrow Q = 330\text{N}$
(c) $P = 330 \text{ N} + 250 \text{ N} = 580 \text{ N}$

14 (a) 20 MJ   (b) 200 J; 2000 m

15 Equate KE to work done by braking force   83 m

16 600 m

17 Each second there has to be a drop in grav. PE of 600 MJ, giving a mass of $1.2 \times 10^6 \text{ kg}$, so mass flow rate $= 1.2 \times 10^6 \text{ kg s}^{-1}$

18 $5 \times 10^4 \text{ N} \times 30 \text{ m s}^{-1} = 1.5 \text{ MW}$
$1.5 \text{ MW} \times (40/30)^3 = 3.6 \text{ MW}$

**Chapter 2
Thermal physics
and states of
matter**

1 $\{(22.4 - 13.2)/(41.0 - 13.2)\} \times 100\,^\circ\text{C} = 33.1\,^\circ\text{C}$ on mercury-in-glass scale
$\{(5.47 - 4.86)/(6.75 - 4.86)\} \times 100\,^\circ\text{C} = 32.3\,^\circ\text{C}$ on platinum resistance scale

2 $2.1 \times 10^3 \text{ J s}^{-1} \times \Delta t = \{(1.2 \times 4200) + 390\} \text{ J K}^{-1} \times (100 - 16) \text{ K}$
$\Rightarrow \Delta t = 220 \text{ s}$
$2.3 \times 10^6 \text{ J kg}^{-1} \times \Delta m = 2100 \text{ J s}^{-1} \times 600\text{s} \Rightarrow \Delta m = 0.55 \text{ kg}$

3 (a) 1.2 MJ   (b) 67 W

4 (a) any slower would make the shower too hot
(b) $\Delta T = 19 \text{ K} \Rightarrow$ outlet temperature of 34 °C

5 E_k of car $= 450 \text{ kJ} \Rightarrow \Delta T = 50 \text{ K}$

6 efficiency $= (1100 - 300)/1100 = 0.73 (73\%)$
$Q_{in} \times 0.73 = 100 \text{ kW} \Rightarrow Q_{in} = 137.5 \text{ kW}$
$Q_{out} = Q_{in} - 100 \text{ kW} = 38 \text{ kW}$

7 (a) 230 W m^{-2}
(b) $\Delta m/\Delta t = 230 \text{ W m}^{-2}/3.3 \times 105 \text{ J kg}^{-1} = 7.0 \text{ kg s}^{-1} \text{ m}^{-2}$
(c) $\Delta x/\Delta t = 7.0 \text{ kg s}^{-1} \text{ m}^{-2}/920 \text{ kg m}^{-3} = 7.7 \times 10^{-7} \text{ m s}^{-1}$
$= 2.7 \text{ mm h}^{-1}$
(d) the rate decreases as the thickness of ice increases

8 $4.5\ \text{W m}^{-2}\ \text{K}^{-1} \times 1.7\ \text{m}^2 \times 15\ \text{K} = 110\ \text{W}$

9 (a) $1.0 \times 10^{-30}\ \text{m}^3$
(b) $10^{-5}\ \text{m}^3/1.0 \times 10^{-30}\ \text{m}^3 = 1.0 \times 10^{25}$
(c) 16.6 mole (17 mole to the required 2 sig. figs.)
(d) molar volume $= 1 \times 10^{-5}\ \text{m}^3/16.6\ \text{mol} = 6.0 \times 10^{-7}\ \text{m}^3\ \text{mol}^{-1}$
$\Rightarrow$ molar mass $= 1.5 \times 10^{-3}\ \text{kg mol}^{-1}$

10 (a) 1200 mol
(b) 34 kg
(c) $22\ \text{kg m}^{-3}$

11 190K (re-arranged equation is $p = (R/M)\rho T$)

12 (a) 480 K
(b) $1700\ \text{m s}^{-1}$
(c) 6.0 kJ
(d) 150 km

13 (a) $430\ \text{m s}^{-1}$
(b) $6.2 \times 10^{-21}\ \text{J}$
(c) $3.0 \times 10^2\ \text{K}$
(d) speed depends on temperature only, so no change

14 (a) stress $= 1.0 \times 10^9\ \text{Pa}$ (1.0 GPa); strain $= 5.3 \times 10^{-3}$
(b) 2.1 mm

15 Extra load = 9400 N; extra stress = 30 MPa; extra strain $= 1.6 \times 10^{-4}$;
extra length = 0.16 m

16 (a) $220\ \text{N m}^{-1}$
(b) using $1/2\ kx^2$, $x = 19$ m
(c) 31 m
(d) $25\ \text{m s}^{-1}$

**hapter 3
urrent electricity**

1 $2.5 \times 10^{-6}\ \text{m s}^{-1}$

2 Molar volume $= 7.2 \times 10^{-6}\ \text{m}^3\ \text{mol}^{-1}$;
there are 6.0×10^{23} atoms in $7.2 \times 10^{-6}\ \text{m}^3$, giving 8.4×10^{28} atoms per m^3.

3 $10\ \text{m s}^{-1}$

4 (a) 1400 C
(b) 17 kJ

5 (a) 540 W
(b) 76%

6

	3Ω	6Ω
series		
($I = 1.3$ A; use I^2R)	5.3 W	10.6 W
parallel		
(use V^2/R)	48 W	24 W

7 (a) 1.2 MW
(b) 47 A

8 $1.1 \times 10^{-6}\ \Omega\ \text{m}$; 0.27 m

9 Area is reduced by 4 times and length doubled so resistance is 8 times larger, 96 Ω

10 (a) 13 Ω
(b) 0.79 A
(c) 7.9 W

11 (a) 0.9 A

(b) 3.3 Ω

(c) 10.8 W

(d) 8.1 W

12 Q and R in parallel are equivalent to 2 Ω so V_{XY} = 2 V

Depends only on ratios, so unchanged.

13 (a) Potential rises from near zero to near +6 V

(b) dark: 0.12 V; light: 5.6 V

14 (a) 2.2 V

(b) 3.8 V

(c) 380 Ω

The LED is highly non-ohmic: its resistance is heavily dependent on operating conditions.

15 The symbols for quantities used here represent *numerical values only*. A final statement is needed after the calculation to return to the actual physics of the circuit.

Kirchhoff's 1st law:

$I_3 = I_1 + I_2$

2nd law to the LH loop:

$(I_1 \times 1) + (I_1 + I_2) \times 2 = 6$

2nd law to RH loop:

$(I_2 \times 3) + (I_1 + I_2) \times 2 = 3$

These reduce to:

$3I_1 + 2I_2 = 6$

$2I_1 + 5I_2 = 3$

Solving these simultaneously:

$I_2 = -0.27$

$I_1 = +2.18$

So the 2 currents are 0.27 A and 2.18 A but the direction of I_2 has been guessed wrongly.

16 (a) Y since it has the larger p.d. for the same current.

(b) X since it has the larger current for the same p.d.

(c) 1.3 V + 2.0 V = 3.3 V

(d) 0.20 A + 0.27 A = 0.47 A

Chapter 4
Oscillations

1 (a) 0.2 m

(b) $10 \text{ s}^{-1}/(2\pi) = 1.6$ Hz

(c) 20 N m^{-1}

(d) 0.4 J

(e) −0.13 m

(f) +13 m s^{-2}

2 (a) amplitude is 5 m and mean depth is 10 m

(b) the equation shows the values in (a) with $\omega = (2\pi/12.5)$ h^{-1} =0.50 h^{-1}

(d) substituting h = 7.5 into the equation gives:

$\cos(0.50t) = 0.5$, so working in radians

$0.50t = \pi/3 \Rightarrow t = 2.09$ (note that t has a numerical value only)

So the earliest time is 2.09 hours after midday, 2.05 pm.

3 (a) 960 m s^{-2}

(b) $\omega^2 = 1.92 \times 10^5$ s$^{-2} \Rightarrow f = 70$ Hz

(c) 2.2 m s^{-1}

4 $\omega = 2.5 \times 10^5$ s^{-1}; $v_{max} = 0.5$ m s^{-1}; $a_{max} = 1.3 \times 10^5$ m s^{-2}

This acceleration could create stresses comparable to the breaking stress.

5 (a) 500 N m^{-1}

(b) 0.03 m

(c) 0.08 m

(d) $\omega^2 = 125$ s^{-2} $\Rightarrow$ $t = 0.6$ s

(e) 0.2 J

6 Possible values: mass = 1000 kg; sag = 1 m
So stiffness in the middle could be about 1×10^4 N m^{-1} giving a frequency of 0.5 Hz. This is 1 cycle every 2 seconds which is comparable to an elephant's step rate, so watch out!

7 f is proportional to $1/\sqrt{m}$ so $f\sqrt{m}$ is a constant
New mass = 64.21 kg $\times \sqrt{\{0.4814/0.4709\}} = 64.37$ kg

8 (a) 1.4×10^4 N m^{-1}
(b) 6 Hz
(c) about 20 m s^{-1}
For the whole car the stiffness is 4 times that for 1 spring, giving a frequency of about 1.2 Hz. The speed to be avoided is about 4 m s^{-1}.

Chapter 5
Waves

1 (a) and (b)

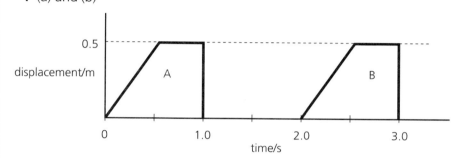

(c) Notice that the pulse shape on the time axis is reversed from that on the original displacement axis.
(d) A moves 0.5 m in 0.5 s so has velocity 1 m s^{-1}.

2 Remember that 1 N is 1 kg m s^{-2}; 7.0×10^{10} Pa (70 GPa)

3 (b) See answer to Ch. 2 Qn.11
(c) 340 m s^{-1} $\times \sqrt{(200/300)} = 280$ m s^{-1}

4 (a) $\sin^{-1}(1.45/1.47) = 81°$
(c) The maximum distance travelled is the length of the cable increased by a factor $(1/\sin 81°)$, i.e. 1 km $\times$ 1.014 $\Rightarrow$ extra distance is 14 m.

5 (a) 11 fringes counted gives 10 fringe spacings.
Direct substn. into the double slit relation gives $\lambda = 5.4 \times 10^{-7}$ m which is in the green part of the spectrum.
(b) Slit separation narrower, observer distance longer, colour towards the red (longer wavelength) end of the spectrum.

6 (a) The path difference between the 2 routes changes, producing changing phase differences.
(b) path difference in first position is $(4 + 4)$ m $-$ 5 m = 3 m
path difference in second position is 8.4 m $-$ 5 m = 3.4 m
this increase of 0.4 m *could be* $\lambda/2$ ($\lambda = 0.8$ m)
but it could also be $3\lambda/2$ ($\lambda = 0.27$ m) or $5\lambda/2$, etc.
It is essential to know whether or not there are any intermediate maxima or minima.
(NB: It is **extremely dangerous** to rely on just one position like (a) in the question diagram to decide on the path difference–wavelength relation. This is because:
(i) you don't normally know the number of whole or half wavelengths which contribute to the path difference
(ii) there may well be an additional phase difference introduced by reflection.
The safest approach (experimentally and in questions) is to say that the *extra* path

length introduced in going from one kind of interference (constructive or destructive) to the *next position of the same kind* is always one wavelength.)

7 (a) 0.80 m
(b) 256 Hz
(c) 2nd harmonic is 512 Hz: λ in air is 0.64 m

8 The wavelength is 1.5 km so a standing wave pattern will be set up along the line joining the transmitters with nodes 750 m apart (about 25 seconds intervals at driving speed).

9 $\lambda = 0.86$ m
1st length $= \lambda/4 = 0.21$ m; 2nd length $= 3\lambda/4 = 0.64$ m
End corrections (where the 'acoustic length' is a bit longer than the physical length of the pipe) have been ignored.

10 (a) 1.7×10^{-6} m
(b) red 23°; blue 14°
(c) 51° and 29°
(d) a 3rd order would give $\sin\theta > 1$ so the highest order is the 2nd.

Chapter 6
Gravitational
and
electric fields

1 (a) See answer to Ch.1 Q.3 for method
force $= (0.5 \times 10^{-3} \times 9.8)$ N $\times \tan 5° = 4.3 \times 10^{-4}$ N
(b) $E = F/q$ and $E = V/d \Rightarrow V = Fd/q = 1.1$ kV

2 (a) 3.3×10^5 V m^{-1}
(b) electric field is 3.3×10^5 N C^{-1}
(c) 5.3×10^{-6} N
(d) Constant force if it is 'well inside', becoming weaker as it approaches the edges; movement perpendicular to the plates has little effect.

3 $q = 4.9 \times 10^{-19}$ C so 3 electrons surplus.

4 (a) 0.22 nF
(b) 2.7 nC
(c) 42 nC m^{-2}
(d) 4800 C F^{-1} m^{-1} or 4800 V m^{-1}
(e) 12 V/2.5 mm = 4800 V m^{-1}
The last 2 results are identical – look at the theory of the parallel plates

5 $\Delta q = 2.4$ nC
$\Delta t = 2 \times 10^{-5}$ s
$I = 0.12$ mA

6 Energy = 160 kJ, giving a charging voltage of 890 V.

7 (a) Gradients are approximately 10, 3.4 and 1.2 μC s^{-1}.
These are the currents flowing at these times (in μA).
They are approximately in the same ratio at equal time intervals, so the decay appears to be exponential (the charge values are 100, 38 and 13 μC approx. which confirms this).
(b) $R = V/I = 10$ V/(10 x10^{-6}) A = 1 MΩ
(c) RC is approx. 10 s, so R = 10 s/10 μF =1 MΩ
(d) (i) starting charge halved; time constant halved
(ii) stays the same

8 (a) 3.3×10^{-9} J
(b) 20 GV
(c) 20 MV
(d) It is moving so close to the speed of light that its relativistic mass increase is significant – data tables give the rest mass.

9 0.15 m

10 (a) $M_e = gR^2/G = 6.0 \times 10^{24}$ kg

  (b) $\rho = 5500$ kg m^{-3}

  (c) Only a mean density; expect lower density material to have 'floated' to surface.

  (d) $g = 4\pi G\rho R/3$, so $g \propto R$

11 6×10^{13} N kg^{-1}

12 (a) 2π radians per year or 2.0×10^{-7} rad s^{-1}

  (b) $M_s = \omega^2 d^3/G = 2.0 \times 10^{30}$ kg

13 (a) $V = -GM_e/r$ (the relation $GM_e = -gR_e^2$ is a useful one to use – but watch the signs!)

  at 10 Mm $V = -40$ MJ kg^{-1}

  at 20 Mm $V = -20$ MJ kg^{-1}

  (b) $\Delta Eg = 800$ kg $\times \{-20 - (-40)\}$ MJ kg^{-1}

  $= +(800 \times 20)$ MJ $= +16$ GJ

  (c) Initial KE = 3.6 GJ

  Final KE = (3.6 + 16) GJ = 19.6 GJ $\Rightarrow$ speed = 7.0 km s^{-1}

14 Use $(1/4\pi\varepsilon_0)\, Q_1Q_2/r$ with charges $+37e$ and $+55e$

  4.7×10^{-11} J (290 MeV)

1 (a) 0.01 N m^{-1}

  (b) 0.09 N m^{-1}

  (c) zero

2 28×10^6 s^{-1} (this frequency of 28 MHz is called the cyclotron frequency and would be the rate of switching voltages between the halves of a cyclotron).

3 2.9×10^{-20} kg m s^{-1}

4 Induced emf = 0.23 V giving a current of 45 mA.

  The connection should be stationary with some sliding contact arrangement, otherwise if the whole circuit moves there is zero emf.

5 (a) 0.4 Wb

  (b) 0.2 Wb

  (c) (0.4 – 0.2) Wb/0.5 s = 0.4 V

6 (a) 1 A s^{-1}

  (b) 7.5 mH

  (c) Steadily withdraw it in 2.5 s. The emf is in the opposite direction. 1.9 s

7 (b) $V_L + V_R = 12$ V

  $LdI/dt + IR = 12$ V

  (c) $I = 0$ and so $dI/dt = 12$ V/ 2.5 H = 4.8 A s^{-1}

  (d) I increases and so dI/dt decreases

  (e) Show initial gradient of 4.8 A s^{-1}, final value of current 1.3 A, gradient gradually decreasing

  (This circuit has a time constant of L/R (0.3 s) so the final current would be effectively reached in about 1.2 s.)

8 24 mA

  There will be two perpendicular fields of 1.8×10^{-5} T giving a resultant of 2.5×10^{-5} T at 45° to the solenoid axis.

9 Use $B = \mu_0 NI/2r$, giving 29 mA.

10 1.3×10^{-4} T

  (b) 1.3×10^{-3} N m^{-1}

11 (a) 1.7 mC (remember the $\sqrt{2}$)

  (b) 141 kV

12 (a) (i) p.d. varies sinusoidally between ± 15 V

(ii) current varies sinusoidally between ± 3 A

(iii) power varies between 0 and +45 W at a frequency *twice* that of the supply voltage (power is always positive)

(b) (i) 2.1 A

(ii) 22.5 W

13 $V_{rms} = 3.7$ mV so $I_{rms} = 1.1$ mA

14 $I_0 = 18.4$ A; $\omega = 100\pi$ s$^{-1} = 314$ s^{-1}

$I = 18.4 \sin(100\pi$ s$^{-1} \times 2.5 \times 10^{-3}$ s$)$ A

$= 18.4 \sin(\pi/4$ rad$)$ A

$= 13.0$ A

15 (a) 2.0×10^5 A

(b) 2000 MW km^{-1} (nearly all the power generated!)

(c) 5500 A

16 (a) 3.2 kΩ

(b) at 100 Hz the capacitor has 1.6 kΩ reactance so this must also be the reactance of the inductor (remember they subtract); using ωL, $L = 2.5$ H

(c) A resonance response graph peaking at 100 Hz

(d) The reactance is zero, so $I_{rms} = V_{rms}/R = 0.5$ A

Chapter 8
Nuclear physics

1 (a) $r = 5.6 \times 10^{-15}$ m; volume $= 1.8 \times 10^{-43}$ m^3

(b) Molar volume $= 7.1 \times 10^{-6}$ m^3 mol^{-1}, so volume of 1 atom is this value divided by the Avogadro constant: volume $= 1.29 \times 10^{-29}$ m^3; ratio is 6.5×10^{13}

(c) nuclear density is this factor $\times$ 8900 kg m^{-3} since practically all the mass is in the nucleus; density $= 5.8 \times 10^{17}$ kg m^{-3}

(d) 9.4 km

2 (a) Since the uranium nucleus is much more massive (look out for clues in the data) its motion can to a good approximation be ignored.

The collision is elastic ('heat' is meaningless on the scale of individual particles – a single atom cannot have a temperature).

(b) KE of alpha particle $= 1.1 \times 10^{-12}$ J; using $(1/4\pi\varepsilon_0)Q_1Q_2/r$ gives a value for r of 3.9×10^{-14} m

(c) For the same energy a proton can reach half this distance since it carries half the charge and electrostatic potential energy α $1/r$

(i.e the ratio Q/r must be the same for proton and alpha for the same energy).

4 (b) Both the background activity and the activity of the source itself will show statistical fluctuations so you should never expect a completely smooth curve (remember that the derivation of the exponential law of radioactivity is itself statistical, relying on a probability argument).

(d) Gradient of best straight line is about -0.013 s^{-1}

0.013 s$^{-1} \times T = \ln 2 \Rightarrow T = 53$ s

5 Convert from half life to λ and use the law of decay $dN/dt = -\lambda N$; 9.3×10^5

6 (a) 2.6 MeV is 4.2×10^{-13} J; 20 mJ per second are needed so the number of decays per second is 4.8×10^{10} s^{-1}

(b) Following the answer to Q.5, $\lambda = 4.1 \times 10^{-9}$ s^{-1}; $N = 1.2 \times 10^{19}$

(c) 1.2×10^{-6} kg

(d) After 3 half-lives the activity would have reduced to 1/8 of its initial value and so the power would then be 2.5 mW. The effective lifetime is therefore about 16 years.

7 (a) Mass of potassium nucleus is 40 u and the result follows

(b) Use the law of decay as earlier; half life is 1.2×10^9 years

(c) The total mass remains constant, so initial mass of potassium must be 6.4 μg.

The mass of the *potassium isotope* has reduced by a factor of 8 which corresponds to 3 half-lives, 3.6×10^9 years

8 (a) 3.51×10^{-3} u $= 5.8 \times 10^{-30}$ kg (keep all the significant figures in until the end otherwise rounding errors can throw away everything that is important)

(b) (i) 3.3 MeV per fusion (use $c^2 \Delta m$ and convert J to MeV)

(ii) 7.8×10^{13} J kg^{-1} (remember 2 hydrogen nuclei are involved)

9 (a) $\Delta m = -0.19434$ u so following answer to Q.8(b) the reaction will produce 2.9×10^{-11} J per fission

(b) 180 MeV per fission

(c) Taking the mass of a uranium atom as 235 u (not quite justified in practice but it is the only indication to go on in the question), 7.4×10^{13} J kg^{-1}

Chapter 9
Quantum effects

1 (a) (i) 3.14×10^{-19} J

(ii) 1.96 eV

(b) 2.2×10^{15} s^{-1}

(c) 2.4×10^{12} m^{-3} (rather like '$I = nevA$' for electrons in a wire)

2 (a) 5.0×10^{-7} m (follow the route eV $\rightarrow$ J $\rightarrow$ f $\rightarrow$ λ)

(b) (i) $E_k = 3.1 \times 10^{-19}$ J $\Rightarrow v = 8.3 \times 10^5$ m s^{-1}

(ii) 1.9 V

3 3.4×10^{14} Hz; in the infra-red ($\lambda = 9 \times 10^{-7}$ m)

4 (a) (i) 3.64×10^{-19} J

(ii) 2.27 eV

(b) level 5 to level 3

(c) $\lambda = 2.52 \times 10^{-7}$ m giving a photon energy of 4.93 eV, a transition from level 2 to level 1 (ground state)

(d) (i) Some mercury atoms are excited from the ground state to n = 2 by exactly the correct incoming photon energy

(ii) The intensity is strongly reduced by the absorption of the photons

5 (a) 1.9×10^6 m s^{-1}

(b) 79 000 K

At much lower temperatures there will still be a few with the required energy – remember there is a very wide distribution (the Maxwell-Boltzmann distribution) about the mean energy.

6

	electron	proton
momentum/N s	3.8×10^{-23}	1.6×10^{-21}
wavelength/m	1.7×10^{-11}	4.1×10^{-13}

7 (a) using the double slit expression $\lambda = 1.5 \times 10^{-10}$ m

(b) 670 m s^{-1}

8 (a) 1.7×10^{24} kg m s^{-1}

(b) 1.5×10^{-18} J

(c) This mode has the largest wavelength and so the values represent minima for momentum and kinetic energy

(d) -2.3×10^{-18} J

(e) -0.8×10^{-18} J: sign indicates the electron is bound to the nucleus – a stable arrangement

(f) E_k is 100 times larger

E_p is 10 times larger (more negative)

so the overall energy is now positive and the electron could not exist at this distance in a stable state.

Index